Secrets of the Seven Metals

A Bridge between Heaven and Earth

Nicholas Kollerstrom

Published in 2014 by New Alchemy Press
www.newalchemypress.com

Copyright © 2013 Nicholas Kollerstrom M.A. Cantab, Ph.D.

The author has asserted his moral right to be identified as the author of this work.

All rights reserved.

Any part of this publication may be reproduced or utilized, even without asking permission of the publishers: but kindly acknowledge the source.

ISBN 9780-9572-79926

Enquiries should be addressed to the publisher.

Printed in Great Britain.

Graphics-editing: *Gimp*; Adobe's *Photoshop Elements*

Other New Alchemy Press books by the author are *Venus the Path of Beauty, Farmer's Moon, Eureka!* and *Interface: Astronomical Essays for Astrologers.*

Contents

Foreword
1. Astral Portraits of the 7 Metals — 1
2. Seven Notes on a Scale — 13
3. Gold, Silver & Mercury — 33
4. Copper, iron, tin, lead — 48
5. Mutations of Mercury — 65
6. Aphrodite, Copper and Venus — 87
7. Aurora, golden dawn-goddess — 96
8. Chemical History and Alchemical Myth — 111
9. When Alchemists made Gold — 127
10. The Kolisko Experiments — 139
11. Plutonium, Pluto's Element — 152
12. Metallic Moments — 159
Epilogue — 176
Appendices — 184
Bibliography

FOREWORD

There are moments in the solitude of night when gazing up into the brilliance of the heavens one is confronted with the vast and infinite mysteries of space, of the coruscating constellations of stars and the ordered procession of the planets. But such moments of contemplation are rare these days. As the sprawling cities with their perpetual illumination have dimmed the silent beauty of the night, so it seems our perceptions have also dimmed, having become clouded and unsure. No longer attuned to the rhythms and cycles of the planets and their influence upon the earth we have become cut off from a broader, more encompassing vision of the universe.

The resulting sense of loss and separation has only been intensified by the jejune worldview of modem science, despite the promise of sophisticated technology. Having stripped the planets of their mythological meaning and magical potency, it has given us instead mere aggregations of lifeless matter, inert spheres spinning meaninglessly through a vast emptiness.

Yet there is a growing disenchantment with this cold and mechanistic view of the universe. The phenomenal growth of interest in astrology, for example, vividly demonstrates the need within people to reestablish their relationship with the cosmos, a need no longer satisfied by abstractions of physicists or the blind doctrines of religion.

Here in this book the foundation for such a new body of knowledge is firmly laid. Here is an extraordinary conjunction of two worlds, that the author is well qualified to make being both an astrologer having earned a Cambridge science degree. Effortlessly he straddles both worlds, validating the knowledge of the ancients by using the techniques and discipline of modem science.

Astrologers have known for centuries that the planets influence earthly events and human behaviour - planting by the moon is as ancient as farming (and was the title of a previous work by the author). Science has never accepted such assertions, but here in this book the reader will see images captured in the careful crystallization of metallic salts, during such planetary aspects as conjunctions and oppositions, which describe the character of the planets involved. The 'signature,' to use an old alchemical tearm, of the planet is imprinted upon its metallic representative on earth. We realise, as the experiments unfold, that an invisible relationship really does exist between the planets and the metals they were said to rule in ancient times. It is this direct perception through our own senses which restores meaning to an otherwise fragmented and disparate world.

The author has researched and lectured widely on this subject and other topics of a Hermetic nature for the last 20 years. This present work grew from a small book entitled *Astrochemistry* published in 1984, but here the theme is greatly expanded with the inclusion of many articles, and woven into a dynamic presentation of these ideas.

The Alchemical doctrine of the seven noble metals is revitalised and transported into the twentieth century with the addition of the radioactive metals uranium and plutonium, and the extraordinary synchronicity of their creation in modem laboratories with the powerful influence of the 'new ' trans-Satumian planets, Uranus and Pluto. But there is another vital aspect the author includes, one that is completely ignored by modem scientific research, and that is the planetary influence upon the psyches of eminent scientists at the moment of their key discoveries. Thus the reader will realise how profound is the relationship between the cosmos, the earth and the individual, how intimately we are all involved with the cosmic process of creation.

It is perhaps fitting here to thank a little-known but brilliant German scientist named Lilly Kolisko who spent her lifetime researching 'the influence of the stars upon earthly substance '; it was her experimental work earlier this century that gave impetus to the research of the present author and his colleagues. While researching my book *Metal Power* it was a humbling experience to come across her work, I had no idea it was being continued by contemporary researchers until after my book was published and an unexpected connection was made with Nick Kollerstrom. Since then it has been a privilege to see *Metal-Planet Affinities* take form. Here is a study of celestial influence that can no longer be swept under the carpet by those who refuse to acknowledge the subtle workings of a living universe. The old world view is rapidly metamorphosing into a vibrant, living exemplar for the future. This book is a potent catalyst for the transmutation.

Alison Davidson
Editor of 1993 edition, co-founder of Borderland Science Research Institute, CA and author of *Metal Power*

1. Astral Portraits of the Seven Metals

A FINE STATUE OF MERCURY, god of healing and thieves, stands in the British Museum, as it earlier stood in a temple of Roman Britain, in the West country. Around this statue were found many rolled-up pieces of lead foil, inscribed with curses – a fact which sadly suggests that early Britons failed to apprehend the nature of Hermes/Mercury. Had blessings been deposited, inscribed on silver, then maybe Hermes could have done something about them. The present work is *Hermetic*, and let us hope that it will not likewise be approached with thoughts of leaden inertia, which will fail to comprehend its arguments. By 'Hermetic' we mean, requiring a certain mobility of thought somewhat akin to the mutable metal mercury, to follow threads of argument woven between the earthly and celestial realms.

This work has been written for astral philosophers, as a series of essays, mainly in astrology journals, somewhat recast for publication. It is concerned with that which really and archetypally exists, and these archetypes *are beautiful.* I say that it concerns the Seven Pillars of Wisdom. A more civilized society would surely wish to apply this Hermetic knowledge in diverse ways. But here today, I cannot guarantee that this knowledge will be useful. For several centuries the Western world has developed concepts that were utilitarian, whereby Mother Nature has been exploited, raped and generally pulverised, treated her as non-living. These inorganic, mechanistic concepts have been tremendously successful in the manipulative purposes for which they were designed. The old alchemic cycle of becoming, including putrefaction, death and rebirth has been replaced by technology's straight line of production-use-dump it.

Thereby we stand on the brink of annihilation.

Suppose we in fact inhabit a living universe, how would things be different? We would need the faculty of *wonder* to appreciate the being, the being-ness of things as a starting-point, and perhaps the ostensibly rather boring realm of inorganic chemistry is a good place to start. This work may have no higher aim, than to enable the reader to gaze at the swirling green hues of a crystal of malachite and really *see* it, to move from the prosaic realm of number, dimension and weight to the qualitative realm of essence and being; and thereby become capable of wonder. I hope it won't sound unduly pious to suggest that the word 'consider' may reacquire its celestial reference of *con-sidera*, 'with the stars', as theory becomes *theoria*, a contemplation of divine principles, as inherent in the realm of matter.

My earlier work *Astrochemistry* presented evidence for a belief which I claimed was older than chemistry or astrology, or even the earliest alchemic texts: the notion that certain metals have an inherent link or affinity with their parent planets.[1] Through many centuries, people were once accustomed to take for granted a living connection between the Earth and the heavens, and this was an expression of it.

The language needed to describe the concepts involved is, perhaps unavoidably, Mediaeval and prescientific. We quote from an essay concerning the great alchemist Paracelsus:

> As it was with *Sol* and gold, so it was also with the other metals and their planets. The metals had somehow, the same *virtu* as the planet, or rather, a single spirit infused both planet and metal, one a celestial and one a terrestrial manifestation of the same force. This was not symbolism, but something much closer to literal speech, a fine line between the two which has been lost to us, and to our language.[2]

In 'Astrochemistry' I described a series of controlled experiments whereby patterns formed by metal ions in solution rising up into filterpapers, responded to particular celestial events. These 'time-experiments' as I call them, over conjunctions and oppositions of the planets, appeared to show that the metal salt solutions were responding to the planets traditionally associated

---

[1] N.K,*'Astrochemistry, A Study of Metal-Planet Affinities*, 1984. For a review see online *Cosmic Influence on Humans, Animals and Plants* 1997, by professor T. Burns; on the same page see his review of my follow-up *The Metal-planet Relationship: A Study of Celestial Influence* (1993) and the article 'Planetary Influence on Metal Ion Activity' *Correlation* 1983.

[2] 'Paracelsus: An Appreciation' by Diane di Prima, in *The Alchemical Tradition in the Late Twentieth Century* Ed. R.Grossinger, Berkeley 1983, p.31.

with them. Reaction rates altered, as did colours appearing on the filterpaper pictures and measurable quantities of metallic precipitate, at the appropriate celestial moments.

Chemical experiments were thus validating the astrological notion of aspects, and the astrological concept of rulership, and also indeed casting light upon the astrological concept of orb. But, these experiments done in my youth have not been replicated for quite a long time, so I've relegated a quite brief account of them to an Appendix (There is quite a bit on the web for those who are curious). The procedure was developed by Frau Elizabeth Kolisko, in the early decades of this century, following indications by Dr Rudolf Steiner. What Kolisko discovered from a chemical point of view - for the benefit of chemists reading this, and I hope there will be a few - was that a reaction involving the precipitation of silver by iron was slow. It was a colloidal precipitation reaction which took several minutes. It is rare within the realm of inorganic chemistry for an ionic reaction to be slow in this manner, normally inorganic reactions are instantaneous.

I haven't come across any chemistry book which discusses or even mentions this simple inorganic reaction as being slow - and

that is quite apart from Kolisko's application of it.

How it Works

As to how these experiments work, or why they should, maybe the great astronomer Johannes Kepler was on the right track when he declared, as regards how the Earth as a whole responded to celestial influence: 'Earth has a vegetative animal force, having some sense of geometry'. The earth is stimulated by the geometric convergence of rays formed round it. The world-soul is sentient but not conscious. As a shepherd is pleased by the piping of a flute without understanding the theory of musical harmony, so likewise Earth responds to the angles and aspects made by the heavens, but not in a conscious manner.[3]

His Pythagorean notion was that astrology worked by a *musica mundana*. Kepler was the last astronomer of note to take astrology seriously, and perhaps further comprehension of the 'how' of celestial influence should proceed along the lines of his opus 'On the More Certain Fundamentals of Astrology.'[4] It was composed in the form of an introduction to an almanack for the coming year.

Nowadays, explanations are supposed to involve a 'mechanism' to be accepted as 'rational.' Our concept of rationality very much involves processes at a microscopic or sub-visible level which produce visible phenomena, and are their explanation. Instead, let us recall the derivation of the term from 'ratio.' Certain symmetrical angles formed between the planets will exert some influence. That is the traditional Pythagorean attitude, twenty-five centuries old, which has an inherent link to musical theory. I find it *rational* to suppose that an angle, such as ninety degrees, forming between two planets, is in accord with effects taking place on Earth, even though no 'mechanism' may exist to account for this.

---

[3] N.K, 'Kepler's Belief in Astrology', *in History and Astrology, Clio and Urania confer* Ed. Kitson, London 1989, p.158.
[4] *De Astrologiae Fundamentis Certioribus*, Prague 1602, translated in 'Johannes Kepler: a Lutheran Astrologer' J.V. Field, *Archive for History of Exact Sciences*, 1984, Vol. 31.

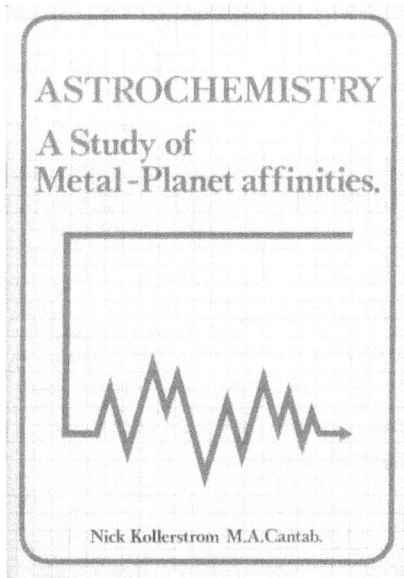 

There was a historic juncture where chemistry and alchemy parted company:

> The microscope marked the end of traditional alchemy and the herbal medicine and *Doctrine of Signatures* that were its allies. With the microscope came a new way of looking into substance, and we have not yet recovered from our curiosity about infrastructure to seek the real centre anew.[5]

What we are here going to mean by 'inner' is definitely not the matrix of atomic structure, but is rather, *quality*. This is not much of a word to bear the freight of meaning we shall be assigning to it, yet it will have to serve.

The opus, *Metal Power* by Alison Davidson[6] has well described Kolisko's experience of the various celestial events she followed by means of her filterpaper experiments. These were primarily qualitative, in that her interest was in the changing images revealed by silver, and indeed also by gold and tin and the various metal salts her experiments used. She made these every day, or sometimes hour by hour depending upon the events involved.

---

[5] Grossinger, op. cit., p.251
[6] Alison Davidson, *Metal Power*, 1992, Borderland Press, CA.

Since Kolisko's time various persons have carried out experiments to check these claims. Some of these results have been negative but most of them were positive. Those described here were performed as carefully as possible, and owed their existence to a research grant provided by the Astrological Association, after the author addressed its annual conference in 1975 on the subject. Most people known to the author who have taken an interest in the experiments have experienced the reality of the metal/planet effect; while their methodology leaves room for improvement. But on the other hand, there is also the feeling that there is so much electromagnetic pollution around these days that the cosmic-etheric influences, however we picture them, are not so easily detectable as they were in Kolisko's day.

This book represents a journey of discovery into the metallic realm. I grew up with a sense of fascination for such chemistry, and used to make fireworks to see how their different hues came from mixing in the various metallic powders. My mother was a chemist and my father was a psychoanalyst, and they argued. I guess this work is a resolution of their opposed viewpoints, as a personality study of the metals.

The reader who embarks on this journey, ranging from the depths of antiquity to outer space, from alchemists trying to make gold to the modern school classroom, may come to share this sense of wonder and mystery, and experience the metals in a new way. The alchemic idea is, that the metals as we know them represent a material expression of living cosmic principles. This was known and directly experienced in times gone by, but then faded away as the more contracted scientific consciousness developed, as the planets became 'mere' objects in space.

While the seven metals known to antiquity have from Greek times been associated with the 'planets' visible to the naked eye, I began to consider whether there was any comparable connection between the atomic-energy metals and extra-Saturnine planets. Initially I doubted whether uranium and plutonium had much of a connection with their celestial namesakes. A breakthrough came on acquiring the exact moment in time when the cyclotron was switched on to create the first sample of plutonium, by Glenn

Seaborg and his team at Berkeley, California.[7] The result in Chapter Eleven was my first attempt to analyse an astrological chart. The time of birth of Glenn Seaborg, whose team had created the unnatural new element was then obtained. I believe that there is a credible link between Seaborg's birth chart and the plutonium natal chart, as described.

Seven Characters

I'm calling this *Astral Portraits of the Seven Metals*, because the seven principal characters of this work are presented in the language of chemistry. These are the primordial archetypes, the metal/planet identities. (The term 'planet' is here being used in its old sense, referring to the seven bodies that can be seen moving across the sky, including both luminaries.) These themes are then further developed in the next section. We appreciate having copper and bronze around for decoration, *because* that helps us to experience its Venusian side.

Becoming aware of the deeper reality of the metals could enrich areas of modern life. For example an artist using metallic copper will better appreciate how to use it if its Venus-nature is comprehended; a doctor applying a gold preparation will be aided by knowledge of its solar attributes; a politician will better appreciate how to cope with the vast issues raised by the existence of plutonium if its chthonic, subterranean or Plutonic nature is appreciated. How sad that people wear gold and silver for decoration, without being aware of their solar and lunar being-essence: this is an empty materialism, of just feeling the greed from owning the gleaming metal.

To-date, the main interest in the matter here presented has been from astrologers, because of the way it validates in a practical way some of the principles of their ancient art. The present work provides a physical basis or counterpart to beliefs that astrologers otherwise experience in a merely psychological context. It is valuable and confidence-building for them to perceive that certain concepts central to their craft operate within the realm of nature, not just in human life.

---

[7] N.Kollerstrom and M. O'Neill, *The Eureka Effect* 1995, Urania Trust.

This work has a lot of relevance to school chemistry lessons, by way of kindling wonder and fascination within the minds of pupils. Let us hope that in the future, school chemistry teachers will wish to use the imaginative truths here presented. A more personally enriching and humanly valuable approach becomes possible when the connection between the intellect and the imagination is fostered: we should not sunder apart the 'bare' facts of chemistry and the rich images of Greek mythology. Today's approach results in many pupils feeling alienated from science. The Mercury chapter does suggest such a pedagogic format, pointing the way to a more wholistic science.

The Mercury chapter was hard to compose, as its theme kept twisting and turning with so many different aspects and loose ends. Years went by while, like metallic mercury, it resisted being fixed into a final form. Then, a neighbour and friend Tony Jackson did an essay on the topic, as part of a homeopathy course. He was a jazz musician and therapist. After his death I got permission from his partner the astrologer Sue Rose to publish it, and I'm sure you will agree, his essay is better than mine!

The next chapter looks at the physiological role of copper in the female organism, relating this imaginatively to the Venus-nature. There was a paucity of references to the vital role of copper-serum blood levels in the female cycle. Textbook discussions on the subject were by men whose interest in copper lay merely in its supportive role for the formation of the iron-molecule haemoglobin. Iron/Mars is normally fairly obvious in the way it manifests, however the more subtle copper/Venus energy in the blood plasma is surely of no less interest. Much of the research on which the copper-chapter is based was carried out at just one American hospital, which specialises in treating patients for mineral imbalances.

Dawn of Modern Science

Years later, and in a different context, I came to investigate some decisive moments in the history of science, which led to the final Metallic Moments section[7]. Events involving specific metals are there considered, where their characters manifest in the event. The theory of correspondences gives us predictions as regards the kinds of planetary alignments we should expect at such times. Further exploration of charts associated with the genesis of atomic energy occurs in this section: is extent uranium 'really' linked to Uranus and plutonium to Pluto? Studying the key moments, when Fermi switched on the first atom pile in Chicago and when Seaborg switched on the Berkeley cyclotron to create the initial sample of the artificial element plutonium, both timed within a minute or so, was an educational experience.

I suggest that it gives us a deeper insight into the history of science to appreciate how these archetypes have worked in its unfolding.

Turning to a different example, I was impressed by the role of the late Carl Sagan in the landing of the Viking spacecraft on Mars, and the formative effect this had upon his life (Chapter twelve). Such events enable us to explore the dynamic relationship between the psyches of some eminent scientists and their historic deeds, through studying the times involved. We look at the configurations of the charts of these men - what an astrologer calls, *transits* - when the key moments arose, as they entered into history through their operations upon matter. (A transit is an event whereby the position of the planets in the heavens come into a relation with one's natal chart). Historic "metallic moments" can often be dated, for example the day when Michael Faraday first generated an electric current using an iron and copper apparatus, but the time is usually lost. I also charted the moment when the poet Alan Ginsberg staged a sit-down protest on a railway line in front of a plutonium-bomb waste waggon and stopped it, having just finished the last stanza of his poem, 'Plutonium ode'. He was just having a Pluto-return (square) but also a Mars opposition-Mars and Moon-trine-Moon: the stars chime, and lend a dignity to human action.

Gold is chemically unreactive, and in consequence there are no notable 'golden moments' in the history of science that can be dated. The source of dated events whereby one can examine the traditional solar nature of gold, lies within the alchemic tradition. A chapter on 'goldmaking' events in European alchemy, aims *not* to raise the issue of whether such events 'really' happened - probably unanswerable - but rather to view them in a philosophical manner as moments in time, with regard to Jung's maxim that 'whatever happens in a moment of time will have the quality of that moment.'

In the seventeenth century, the notion of the 'inside' of things suffered a sea-change, as it ceased to mean a qualitative/symbolic essence, and instead came to signify what could be seen down a microscope. To quote the science historian Alexandre Koyré, the 'scientific revolution' in Europe involved:

> ...the disappearance - or the violent expulsion - from scientific thought of all considerations based on value, perfection, harmony, meaning and aim, because these concepts, from now on merely subjective, cannot have a place in the new ontology.[8]

Plenty have complained about this dilemma, but it remains with us. The psychologist R.D.Laing described how Galileo's vision of a universe described only in terms of mathematics led to the loss of:

> ...aesthetics and ethical sensibility, values, quality, form; all feelings, motives, intentions, soul, consciousness, spirit. Experience as such is cast out of the realm of scientific discourse.[9]

That is quite a lot to lose! If we start from something fairly simple, viz. inorganic chemistry, and try to discern what rightly belongs to the cosmic process, we may hope to make progress.

In Britain, leading illuminati of what later came to be called the scientific revolution favoured a sobered-down and 'purged' or purified astrology. Francis Bacon attacked astrology and alchemy in his *Advancement of Learning*, but then later came to modify his view: the trouble with astronomy, he opined three and a half centuries ago, was that it had been taken over by the mathematicians. The world stood in need, he affirmed, of a very different astronomy, of *a living astronomy*, of an astronomy which should set forth *'the nature, the motion, and the influences of the heavenly bodies, as they really are.'*[10] These are fine and important words from a British philosopher, greatly ignored by those claiming to use 'the Baconian method.' We here foster what Lord Bacon called 'a living astronomy'

Bacon objected to astrology on the grounds that it was 'so full of superstition,' yet said that he 'would rather have it purified than

---

[8] Koyré, A. *Newtonian Studies*, 1965, p.7.
[9] R.D.Laing, *The Voice of Experience*, New York, 1982.
[10] Bacon, F. *De Augmentis Scientarium*, Book 3, Ch 4.

altogether rejected.' He hoped that it would be possible to use the celestial indicators for predicting 'all commotions or greater revolutions of things, natural as well as civil' which phrase quite well applies to the 'metallic moments' we shall be here examining. 'I will add one thing besides (wherein I shall certainly seem to take part with astrology if it were reformed); which is, that I hold it for certain that the celestial bodies have in them certain other influences besides heat and light.'[11]

The architect Christopher Wren, later President of the Royal Society, expressed such a Baconian view in an early 1657 lecture:

> there is a true Astrology to be found by the enquiring Philosopher, which would be of admirable Use to Physick, though the Astrology vulgarly receiv'd, cannot be thought extremely unreasonable and ridiculous...[11]

The view of Robert Boyle, one of the founders of the Royal Society, who came to be called the 'Father of modern chemistry', has a special interest:[12]

> ...it may, notwithstanding all those objections, still be certain, that these celestial bodies, (according to the angles they make upon one another, but especially with the sun or with the earth in our meridian, or with such and such other points in the heavens) may have a power to cause such and such motions, changes, and alterations... as the extremities of which shall at length be felt in every one of us.

Such were the views of some of the early pioneers of science, as have not yet been developed in the course of time.

The essays presented in this volume unfold a theme not easy to express, concerning a connection between earth and sky. At times over thrice seven years I felt like some alchemist tending the Great Work, which could not be hurried, but it has now matured sufficiently to come out of the athenor.

---

[11] Quoted in Curry, P. *Prophecy and Power, Astrology in Early Modern England* 1989, p.61.
[12] Curry, op. cit., p.64. ' *Of Celestial Influences or Effluviums in the Air*' by Robert Boyle was addressed to Samuel Hartlib, and published posthumously in 'History of the Air', 1691, see Curry Ibid. p.63.

Thanks are due to several persons chiefly within the UK's Astrological Association who have supported these endeavours, especially its late President Charles Harvey; to (many years ago) the late John Davy at Emerson College, who was given an OBE for science journalism and who introduced me to the Kolisko experiments in 1970[13]; to R.M. who after all these years still has to remain anonymous and to Michael Drummond, both of whom participated in the Kolisko experiments, only briefly described here, which drew me into the whole topic; and last but not least to my Mother, who had a doctorate in chemistry, and who taught me how to be intrigued by chemical issues.

An alchemical Hermes, where is he going?

---

[13] I wrote to Kolisko, in 1975, and a friend of hers later told me she received my letter, but she never replied. That was I believe the year when she had a great bonfire of her remaining notes saying 'no-one will want these.'

I. SEVEN NOTES ON A SCALE

*Wisdom has set up her temple,
She has hewn her seven pillars -*

Book of Proverbs[14]

From late antiquity up until the mid-eighteenth century, the number of metals known and recognised as such was seven. They were: lead, tin, iron, gold, copper, mercury and silver. Brass, made from copper, was used, but no-one realized it was an alloy that used zinc, until the latter half of the eighteenth century. The metal which finally broke the sevenfold spell of millennia (in 1752) and was called the 'eighth metal' was platinum, emerging from the gold mines of Columbia.

Belief in a linkage of these seven metals with the 'seven planets' reaches back into prehistory: there was no age in which silver was not associated with the Moon, nor gold with the Sun. These links defined the identities of the metals. Iron, used always for instruments of war, was associated with Mars, the soft, pliable metal copper was linked with Venus, and the chameleon metal mercury had the same name as its planet. Then, around the beginning of the 18th

---

[14] Proverbs, 9:1.

century these old, cosmic imaginations were swept away by the emerging science of chemistry. The characters of the metals were no longer explained in terms of their cosmic origins but instead in terms of an underlying atomic structure. New metals started to be discovered which made the old view appear limited.

In the 20th century new lines of approach to this old subject were opened up through work done within the Anthroposophical movement founded by Rudolf Steiner, and we here draw especially from the works of Rudolf Hauschka[15] and Wilhelm Pelikan.[16] They viewed the traditional seven metals as expressing most fully the seven planetary characters, in a way that the many other metals known today do not:

> The seven fundamental metals represent something like the seven notes of a scale. As there exists a great variety of intermediate tones within the scale, so one can recognise intermediate tones between the metals.[17]

The extra-light metal lithium is used for hydrogen bombs, anti-depressant pills and bicycle axle grease, thereby one feels its 'lightness of being.' But, that won't give us a planetary affinity for it. Magnesium emits a brilliant light on burning, used for photo flashlights, so would this make it solar? It is used for ultra-light alloys in supersonic aircraft etc, and is the key metal used in chlorophyll, whereby solar energy is metabolised by plants. Wilhelm Pelikan suggested that it should be viewed as a sun-metal, and let's leave that open as a possibility.

Physical Properties

We experience metals as differing from non-metals by virtue of their lustre, their resonance, their malleability and conductivity - these are their *key physical properties*. Metals can be polished to shine (lustre), will produce tones when struck, ie they *sound* (resonance), when hammered they don't shatter, they can be beaten into shape, and will quickly become hot if one corner is heated. The traditional seven metals can be arranged in a scale, by these key physical properties.

---

[15] Rudolf Hauschka, *The Nature of Substance,* 1966.
[16] Wilhelm Pelikan, *The Secrets of Metals,* 1973.
[17] Walter Cloos, *The Living Earth,* 1977, p.123.

This turns out, remarkably, to be the same scale as an ordering of their associated planets, in terms of their speed of movement. The Table below expresses metallic conductivity both as thermal (conducting heat) and also as electrical, scaled for convenience to silver = 100.[18]

The planets are ordered by something which one can experience quite directly, namely how fast they move across the sky - from the Moon as the fastest moving to Saturn as the slowest. This means using a *geocentric perspective*, as we see their mean angular speeds from the Earth, and gives the traditional ordering as used to be assigned to the planets in the old, Ptolemaic system - for almost two thousand years. This ordering was almost universally accepted, up until the time of Copernicus, and had the sphere of Mercury *nearer* the Earth than Venus (mathematically, this may be the case: ie, Mercury is more often nearer to us than Venus.[19]

---

[18] Kaye and Laby, *Physical and Chemical Constants*, 14th ed.
[19] N.K., '*Interface, Astronomical Essays for Astrologers*', 2013, Ch.3.

Here is an old engraving of the astronomer Claudius Ptolemy, with the seven spheres of the planets behind him, from the furthest-away sphere of Saturn down to Mercury the fastest-moving planet- you can't see the bottom or seventh sphere of the Moon, closest to Earth. In sequence their glyphs are:

$$\saturn, \jupiter, \mars, \odot, \venus, \mercury, \moon$$

People have believed in that ordering of the heavenly spheres for a far longer time than they have accepted the modern world-view.

If you're hoping to get to heaven when you die, you might want to check out Dante's *Paradiso* which gives the identical sequence of the heavenly spheres. Dante first arrives in the Moon-sphere, then that of Mercury, and so on in this same order until the glorious sphere of Saturn is reached.

Table 1: Metallic Conductivity vs Planetary Motion

Planet	Speed	Metal	Conductivity of	
	Deg./day		Warmth	Elec^y
Moon	13.2	Silver	100	100
Merc.	1.4	Merc.	(68) 2	(76) 2
Venus	1.2	Copper	94	95
Sun	1.0	Gold	74	72
Mars	0.5	Iron	20	17
Jupiter	0.08	Tin	16	13
Saturn	0.03	lead	8	8

(scaling electric and thermal conductivities to silver = 100; values in brackets are for solidified mercury)[20]

---

[20] The same values were used both in McGillian 1982 and Geoffrey Dean 1977 in this context, only I've here scaled them up to silver = 100.

We can summarise the linkage we are here looking at, in the words of biochemist Dr Frank McGillion:

> The orbital motion of the planet correlates in sequence with its corresponding metal's conductivity... The slower a planet moves, the less able its corresponding metal is to conduct electricity![21]

For the alchemists of old, metals shared their characteristic properties in different degrees. They were *not* separate elements, and had these *experiential* properties in common. A metal was purified in a furnace, where it would melt but not burn. Zinc could never be a metal, because it quickly burnt up on being heated. These criteria put them in a quandary over mercury: it was generally recognised as metallic, though paradoxically so.

This experiential definition limits us to what we'll call 'real' metals, whereas the modern definition of a metal is wholly abstract - in terms of atoms that are electron-donors - and includes substances that don't at all resemble these: for example, potassium is a waxy substance that bursts into flame upon contact with water. A lump of sodium will buzz around on the surface of water, in a quite exciting manner. But be careful not to cut off too large a lump of this soft, waxy metal.

Nowadays, children in science lessons are given these totally abstract concepts, that will never impact upon their lives, and are hardly allowed to experience the primary properties of everyday metals. But here we concentrate on things that are *elementary*. Thus, contrasting the front and back gardens, we see gleaming steel on the car in the front garden, and the black wrought-iron gate at the back: how strange that these are the same element, with just a few percent of carbon added, and that is the mystery of iron, whereby it will turn into steel.

In an elementary school the Mars-iron stories would be taught first, involving the clash of steel: eg how the ancient Romans kept winning their battles because they could forge iron and steel which their opponents could not. (A modern iron-Mars archetype might be Popeye the sailorman, who ate spinach to become extra-strong, the

---

[21] Frank McGillian, *The Opening Eye,* 1982, p.94. The standard electrode potentials are here given to the most common valence condition.

idea being that spinach was high in iron. It seems that this is not actually the case, it was all a big mistake!)

Here are the key physical properties:

Conductivity: copper is used for electrical wiring being a good conductor, as lead is used for fuses because it is such a poor conductor. Mercury is not included on this table being a liquid - conductivities of metals when liquid are much lower than when they are solid.

Lustre (or reflectance): silver is the most perfectly reflecting metal of the seven and is therefore used for making mirrors. Mercury also has a very high lustre and is likewise used for such: these are the two mirror-metals. In antiquity, mirrors of copper or bronze were used. The other metals show an approximate gradation in lustre down to lead which has a very dull surface.

Resonance: copper is much used in musical instruments because of its high resonance although silver instruments have the clearest, purest tones - 'silver bells', and this property again decreases down the scale to the dull sound lead makes on being struck.

Malleability: Hauschka described how metals at the top of the list are highly malleable, but cannot be well cast, whereas those at the bottom can be cast but not forged. Gold he described as holding a balance position in that it could equally well be cast or forged.

These scales show an increase in inner mobility from lead, the most inert, up to silver, which parallels the increasing angular speeds of the planets. Hauschka, who first described this, concluded memorably:

We see then that planetary movement is metamorphosed into the properties of earthly metals.[22]

Chemical Activity

> *'This isn't just a date, It's ... chemistry*
> from the film, 'Something about Mary'

---

[22] Hauschka, op.cit., p.162.

Valency Valence is the combining ratio: hydrogen has a valency of one, oxygen of two, and carbon, four. It tells how many 'arms' each element has, whereby it joins up with others. One carbon atom bonds with four hydrogens to give methane (CH_4), while oxygen bonds with just two hydrogens, to make water as H_2O.

Most metals have more than one possible valency state. The Table shows the valencies which the seven metals normally display, while any others that can form are rare and unimportant. Oddly enough, their valencies line up with the traditional Ptolemaic ordering of the heavenly spheres.

Metal	Silver	Mercury	Copper	Gold	Iron	Tin	Lead
Valence	1	1 + 2	1 + 2	1 + 3	2+3	2+4	2 + 4
Planet	☽	☿	♀	☉	♂	♃	♄

<u>Table 2 – Valences, i.e. combining ratios</u>

Silver, which showed the highest conductivity and gave the purest sound, has only a single valency for every link it forms with other elements. Like swans which remain monogamous and faithful to one partner all their life, the Moon-metal silver has only one arm of valence. In contrast, those which scored lowest on their physical properties, tin and lead, being least conductive etc, are most active and greedy in their ratios of combination.

***Reactivity*:** Some metals are inert, for example gold hardly combines at all, these are the 'noble' metals (platinum, silver); whereas tin and lead are reactive and will dissolve even in weak acids. We can put the classical metals in a sequence of their chemical activity, which is conveniently measured by what chemists call their 'electrode potential.' This tells us how reactive their ions are in solution. Inactive metals as will not liberate hydrogen from an acid are called 'electronegative', while the more active metals which will liberate hydrogen are 'electropositive'. This gives a useful scale of chemical activity for metals, measured by the 'standard electrode potential' of a solution at a given concentration.

Let's start (as McGillian here advocated) with the order of the planets going out from the Sun, and then the corresponding electrode potentials of the metals are:

Electronegative				Electropositive		
☉	☿	♀	Earth	♂	♃	♄
Gold	Merc.	Copp.		Iron	Tin	Lead
-1.50	-0.79	-0.33		+0.04	+0.14	+0.13

Table 3: Planetary sequence and electrode potentials

Thereby McGillian contrasted the more reactive, 'electronegative' metals as linked to planets *inside* Earth's orbit with electropositive ions which correspond to those *outside* the Earth's orbit.[23] Electrode potential is measured with respect to that of the earth, which indicates the relevance of the geocentric viewpoint here involved. He concluded,

> The earth-centred universe of the alchemists is polarised into positive and negative. It is chemically yin and yang.

A more traditional ordering (Tables 1 & 2) puts silver at the top of the list and Sun-metal gold in the middle. That gives us 'above the Sun' planets, Mars, Jupiter, and Saturn having electropositive metals, and vice versa for 'below the Sun' planets (Silver's standard electrode potential is -0.8). Either way, the correlations are impressive.

Atomic Number

Each element has an 'atomic number', and the Periodic Table of Elements arranges them in sequence: hydrogen has an atomic number of one, carbon 6 and oxygen 8. The atomic weights are roughly double this, so carbon has atomic number 12 and oxygen 16. This ordering by atomic numbers (or weights) gave insight into the chemical properties of each element. Mendeleev discovered how to do this, to arrange them in this way, and it was called the Periodic table. Thereby he found 'gaps' and could predict the existence and chemical

---

[23] McGillian, op.cit., p. 94

properties of several elements not yet found. In the next century, these numbers were 'explained by concepts of atomic structure. But we're not really here concerned with that, because this isn't a science treatise, it's an alchemical one.

Mendeleev's Table has seven rows or 'periods,' from the first row that has the lightest elements, hydrogen and helium, down to the seventh which has the extra-heavy, radioactive elements such as uranium and plutonium. Vertically, it has seven or eight columns (the eighth and last column with the inert gases is usually given as the zeroth column, with the others counted as 1-7): in a sense it also has seven columns. What are called 'group one' elements belong to its first column, and these are all univalent, such as sodium. Group two (the second column) are bivalent like calcium, group three are trivalent, eg aluminium and so on. The number seven appears in this Table as dominant, and controlling the possibilities of what elements can exist.

When Uranus was discovered in 1781, by William Herschel, this kicked out the notion that there was something sevenfold about the heavens. Up until then, there had been seven spheres which could be seen to move across the sky. There still were such, but an extra unseen one had been added. After his discovery, there was no longer anything sevenfold about the world! This dire state of things persisted for nearly a century, until chemistry professor Dmitri Mendeleev formulated his Periodic Table. A seven fold pattern then reappeared in matter, in the science of chemistry. Bearing this in mind, it may be of interest to look at the moment in time when this new synthesis was created: the afternoon of March the first, 1869. There were no less than six *septile-aspects* then present in the sky, between the planets. They were:

☽-♄ (1°), ♀-♃ (1° 10'), ☽-♅ (0° 10'),

☿-♆ (1° 40'), ♀-♆ (0° 30'), ♄-♅ (1°)

The septile is a celestial aspect formed by dividing the circle into seven parts. It gives the angle of slope of the Great Pyramid, 51½°. The cosmos was in quite a sevenfold mode at that moment in time, when the new synthesis dawned upon Mendeleev. He had cut out cards for each known element, was trying to arrange them by their atomic numbers on his living-room carpet, dozed off, and when he woke up, it came to him! A sevenfold pattern was discerned in matter, during a period when sevenfold aspects were strong in the heavens.

Three Heptagons

Sequencing the classical seven metals by their atomic weights gives an order, which derives from our previous ordering using a heptagon pattern. If you put the seven metals in a heptagon *circle* in the sequence of their physical properties, as given above (Tables 1 & 2), then starting from iron, with the lowest atomic weight, and score alternately, which will give the ordering by atomic weights.

This is confusing! So instead, let's here start off with the seven planets in their day-of-week ordering. The French names of the days of the week give us the planetary names in Latin. Thus they are named after planetary deities, and the European languages (except Greek) concur in this respect. Thus Thursday derives from 'Thor's day', while the French Jeudi is 'Jupiter's day', the thunder-wielding Thor being a Norse equivalent to Jupiter. Likewise there is an analogy between our Friday, as 'Freya's day', and Vendredi, 'Venus' day', with Freya as a Venus-deity, and so forth.

Here is a 19th-century bracelet giving this day of week order, see if you can recognise the deities. It starts (wrongly, one might say) from Monday, with the Moon-goddess. Then Tuesday for Mars, Wednesday for Mercury etc, ending with Apollo for Sunday. We may call this the Days of Creation sequence because, as we'll see, it's the same as the sevenfold creation-sequence that unfolds in the first chapter of Genesis.

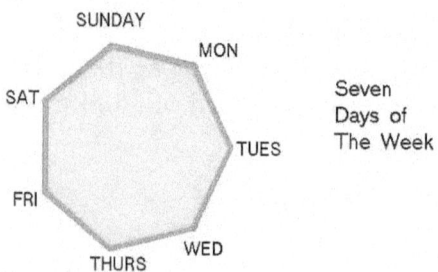

Seven Days of The Week

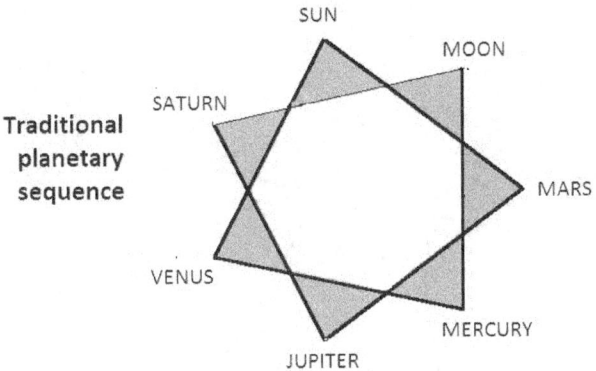

Using that heptagon-circle, select alternately to obtain a bi-heptagon: going twice round that gives you the ancient, Ptolemaic ordering of the planets. Start from the Moon, as the sphere closest to the Earth, and you should end with Saturn as the most distant sphere. We saw how this refers to *both* their speeds of motion across the sky, *and* also to the ordering of the corresponding metals' physical properties.

Take it slowly, one step at a time.

Old books on astronomy described this sevenfold transform, from the Days of Creation sequence, i.e. the seven days of the week, to the old ordering of the planets, and they called this, the 'Hebdomad.'[24] Then, early in the twentieth century, the amazing third step of this argument was discerned,[25] whereby selecting every *third* step around the circle created a star-heptagon, giving the ordering by atomic weight (or atomic number) of the metals! (N.B. Don't confuse this with density, it's not the same). One here starts from iron, as having the lowest atomic weight of the classical seven.

---

[24] C.Leadbetter, *A Complete System of Astronomy*, 1742
[25] Sephariel, *Cosmic Symbolism*, 1912 (quoted by Dennis Elwell in *The Loom of Creation* 2000, as the earliest source he could find).

Secrets of the Seven Metals

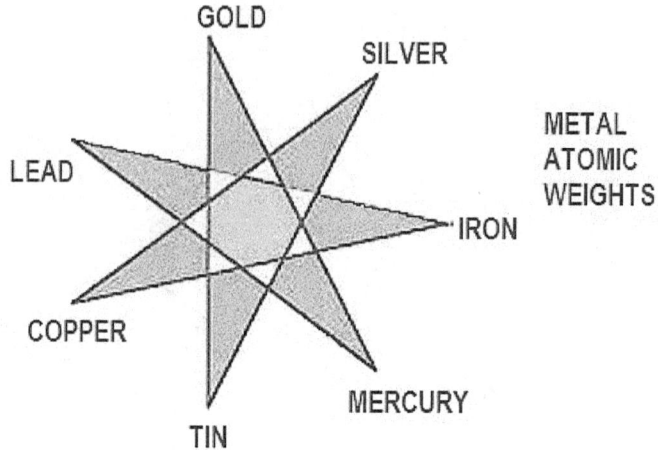

Atomic number follows the sequence of elements in the Periodic Table: hydrogen is 1, helium is 2, etc, with carbon as 6. Atomic weights are roughly double this, eg carbon has an atomic weight of 12. It may help to remember that isotopes come in here: the carbon-14 isotope has a different atomic weight, but it has the same atomic number because it's the same element.

Classical metals	Atomic number
Iron	26
Copper	29
Silver	47
Tin	50
Gold	79
Mercury	80
Lead	82

We don't really need atomic weights for this treatise, it's the atomic number sequence that gives the *magic heptagon* link-up of heaven and earth. Lead has the highest atomic weight of these seven metals – metals of higher atomic weight than lead are radioactive, they cannot remain stable. If you follow the heptagon, the sequence ends with gold, mercury and then lead, with atomic numbers 79, 80 and 82 –

even though gold is very much denser i.e. heavier than either lead or mercury. How strange!

Thus we see how a sevenfold pattern or mandala starts from the names of ancient sky-gods, somehow linked to the days of the week, and then contracts into sequences of physical and chemical properties of the metals. Wilhelm Pelikan was the first to describe these heptagon-patterns, though not in quite the sequence here presented. In a beautiful and mysterious manner, they link together the concepts of modern chemistry and ancient traditions of the cosmos. From a totally unexpected source, we receive confirmation that there is indeed something special about the 'seven metals' known to classical antiquity.

One American academician, Derek de Solla Price,[26] was impressed by the way these heptagons worked, linking atomic weights of the seven metals and the revolutionary period of their respective planets. He was moved to write:

> It seems quite plausible that much of astrological theory may rest on just such a basis of figurate rationality rather than upon empirical or special omen lore. In this sense astrology ... developed on a very rational basis, with a figurative theory and the associated symbolism at its centre.'

Try making your own heptagon, putting the seven planets around it in their proper, traditional sequence.[27] Isaac Newton made a heptagon diagram of the planets in that order, as we will see in Chapter eight. From that, a star-heptagon will takes you into the seven days of the week.

A Sensible Approach

We have here looked at the primary concordances, what one might call the Seven Pillars of Wisdom, from a geometric, heptagonal perspective, to link Earth and Sky, star and stone, psyche and cosmos. Our approach has been *rational,* in the sense of looking at the *ratios* that are involved. Other chapters will outline a more qualitative experience: astral portraits of the metal-planet archetypes. Astrologers,

---

[26] Dennis Elwell, *Astrol. Assoc. Jnl*, review of 'Astrochemistry', Winter 1984/5 p.54.
[27] This heptagon-linkup sequence was given by John Martineau, in his *Little Book of Coincidence:* at www.woodenbooks.com, select this and have a read.

in describing their archetypes, tend to use the old, Greek gods. No doubt these are fine, but an appeal is here made, to seek a more material and experiential basis in the realm of inorganic chemistry. This may seem credulity-straining, but let us see what can be done.

Any answer to the question, 'What is matter made of?' is going to be firmly four-square. The old four-element matter theory came unstuck in the seventeenth century, then reappeared in the twentieth century with the recognition of four states of matter (solid, liquid, gaseous and plasma, the latter being very hot). Then in the 1990s, after a zoo of subatomic particles had appeared, a twelvefold symmetry surprisingly emerged, with six quarks and six leptons, quite analogous to the twelve zodiac signs, with their three families of four. Today, physicists believe that all matter is composed of twelve particles, plus there may be another twelve 'force' particles such as 'gluons' ets. If I've got it right, only four of those twelve matter-particles normally exist, i.e comprise ordinary matter. It begins to look as if high-energy physics (what used to be called, particle physics) requires a grounding in Pythagorean metaphysics, in terms of the significance of the different number-patterns that are turning up.

We are here concerned with *sensible* things, as can be experienced and perceptible to the senses. In contrast, particle physicists are concerned with the *occult*, with what is hidden, as none of the things they deal with can ever be seen. Their particles get smaller and stranger as the budgets grow larger. A British MP visited Geneva, to see the huge underground ring where the particles are accelerated, and emerged claiming to understand what a 'Higgs-boson' was. This was a particle they had recently discovered, which lasts for a millionth of a second or so. Over years, the international project of building these giant accelerators has greatly failed to produce anything the public can understand (Fermilab in the USA just outside Chicago did at least discover the 'top quark').

We are here contrasting the numbers twelve and seven. Twelvefold and fourfold patterns of modern physics concern the very structure of matter; whereas, we saw how sevenfold structures make an earth/sky concordance, in a way that validates the traditional correspondences. These are the seven metals which have been intertwined with the story of humanity. Admittedly, there will be more to say in relation to the outer planets, especially the dire Pluto - plutonium linkup.

The 'Brothers of Iron': Cobalt and nickel, chromium and manganese show a strong affinity with iron, and Hauschka called them the brothers of iron, affirming that they had a Mars-like nature. They have similar properties of resonance and lustre, and iron is hardened by steels having traces of these metals. 95% of manganese production goes towards making steel. Nickel and cobalt both behave like iron when in a magnetic field. The phrase Pelikan used for these metals was, 'we have strong reasons to suspect that the iron-Mars impulse cooperated in their formation.'[28] They are more chemically active than iron,[29] which is why they are more recent.

Praise

Jupiter for Splendour in worldly success
Mars for Bravery in struggling for success
Venus for success in Love
Mercury for the Magic Touch
Saturn for a sense of Destiny
Sun and Moon as alternating
Heavenly Bodies
Parents of our life
On Earth

And Earth
Oh for God's sake
Let's praise
EARTH

       Patricia Villiers-Stuart, Anthem for 2000

---

[28] Pelikan op.cit, p.156
[29] Their divalent electrode potentials are, manganese +1.2, cobalt +0.28, and nickel +0.25. Here are some physical properties of interest:

The 'Brothers of Iron'

Metal	Atomic Number.	Specif. Gravity	Melting-point
Chromium	24	7.2	1900° C
Manganese	25	7.4	1200° C
Iron	26	7.9	1528° C
Cobalt	27	8.5	1524° C
Nickel	28	7.7	1500° C

The Days of Creation

Creation in the Book of Genesis expressed these very same archetypes in day-of-week order. The modern seven-day week with the astrological archetypes embedded into them appeared in Alexandria around the first and second centuries BC. The original Greek text of the Old Testament, called the *Septuagint,* was also created in Alexandria around that time. This was the city where astrology as we know it was born. This opening chapter of Genesis was surely composed there and then. Traditionally that text has been dated around the 6th century BC but it could be more recent.[30]

Hebrew text pre-dating the Greek Septuagint, of the first chapter of Genesis, does not exist. Genesis Chapter Two has a much older text, where God makes Adam and puts him in a garden – this could well be Sumerian, but at any rate it's an older creation-story. It has a male-singular god (Yahweh) whereas the first chapter's creation-story features the Elohim who are plural and do not have gender: the creator-gods.

We've seen that the sequence of the seven days of the week derives from the sequence of the planets, as they were universally known and understood, from Saturn the outermost to the Moon nearest, and we've looked at their lovely heptagons. That happened in Egypt in Alexandria, by a process outside our present concern. It was to do with the idea of 24 hours in a day as developed in Egypt. The day-of-week sequence that was forever after embodied in the cycling seven-day week, was totally determined by this logic, of the 'hebdomad' as astronomers called it, which is basically the heptagon counting we looked at earlier. I here argue that the Creation of the world in seven days in the Book of Genesis reflects and expresses that sequence.

The first day of the week was regarded as being Sunday. For example, The New Testament has: 'Upon the first day of the week, the disciples came together to break bread' (Acts 20:7). , In modern German, Wednesday is called Mittwoch, 'middle of week,' implying that the week starts on a Sunday.

The first Day of Creation, Sun-Day: "*In the beginning God*

---

[30] N.K., *The Days of Creation,* ISAR journal, USA, 2013 (online at 'Astrozero').

created heaven and earth. Now the earth was a formless void, there was darkness over the deep, with a divine wind sweeping over the waters. God said, 'Let there be light' and there was light. God saw that light was good, and God divided light from darkness. God called light 'day' and darkness he called 'night'. Evening came and morning came: the first day."

The first Day is a primary solar image, of the Light appearing, separate from the dark.

There is a wonderful sevenfold affirmation whereby the Elohim keep seeing that each Day is good. This has a lot to do with the forward-moving optimism of the Western world. The world is not made as a mistake, or by some evil or inferior deity. Note that there is no singular deity in this text, 'Elohim' the creator-gods' as here used are neither singular nor masculine. – the male-singular deity Yahweh only turns up in chapter two of Genesis.

The Second Day is amniotic, with water above and below, quite lunar:

> *God said, Let there be a vault through the middle of the waters to divide the waters in two'. And so it was. God made the vault, and it divided the waters under the vault from the waters above the vault. God called the vault 'Heaven.*

This is the one day which is not seen as 'good.' You know that Monday morning feeling? Also the number two is concerned with duality and stress, as the opposition aspect is stressful.

The Third Day, the Tuesday of Creation, is the Mars-day. How might one expect Mars-energies to work in a creation-process? It cannot be fire, for that is destructive; nor war, nor the Smith's forge.

> *God said, Let the waters under heaven come together into a single mass, and let dry land appear.' and so it was. God called the dry land Earth and the mass of waters 'seas', and God saw that it was good. God said, Let the earth produce vegetation: seed-bearing plants, and fruit trees on Earth, bearing fruit with seed inside, each corresponding to its own species.' and so it was. The earth produced vegetation: the various kinds of seed-bearing plants and the fruit trees with seed inside, each corresponding to its own species. God saw that it was good.*

So that which is dry appears, then Earth is inseminated with seed. The

Mars-process here is that process of insemination, of the Earth with seeds. The third Day has the Elohim *twice* seeing that it was good, reminding us of the harmony of a trine aspect. This Day gets a double blessing! Our space-time world first begins to appear on this 3rd Day.

Next, <u>Mercury's Day,</u> has the *signs placed in the firmament, in order that* the festivals can take place here on Earth at their proper times. That interlinking between Earth and sky *is* the Mercury-process:

> *God said, 'Let there be lights in the vault of heaven to divide the day from night, and let them indicate festivals, days and years. Let there be lights in the vault of heaven to shine on the Earth' And so it was. God made the two great lights: the greater light to govern the day, the smaller light to govern the night, with the stars. God send them in the vault of heaven to shine on the Earth, to govern the day and the night and to divide light from the darkness. God saw that it was good.*

Theologians and scientists can discuss *ad nauseam* how come plants appeared on land *the day before* the stars appear in the sky, but they are never going to get anywhere until they (gasp) ask the astrologer, and she will explain the archetypal realities here alluded to: whereby the Mars-day had to come before the Mercury-day. The Mars-process of fertilising the Earth came first, then came the Hermes-Mercury process linking Heaven and Earth, with signs and lights in the heavens to help seasons and festivals to take place.

<u>The Thursday of Creation</u> is expansive and optimistic, when the deep comes to teem with all manner of life, and all is held in balance:

> *God said, 'Let the waters be alive with a swarm of living creatures, and let birds wing their way above the earth across the vault of heaven.' And so it was. God created great sea-monsters and all the creatures that glide and teem in the waters in their own species, and winged birds in their own species. God saw that it was good. God blessed them, saying 'be fruitful, multiply, and fill the waters of the seas.*

There is a grand optimism in the way the birds wing their way across the vault of heaven while huge monsters teem in the deep, and in the divine command to 'be fruitful and multiply'.

<u>Friday, Venus' day,</u> has the man and woman created, 'in the image' of

the Elohim:

> *God said, Let us make man in our own image, in the likeness of ourselves, and let them be masters of the fish of the sea, the birds of heaven, the cattle, all the wild animals and all the creatures that creep along the ground. God created man in the image of himself, in the image of god he created them, male and female he created them. God blessed them, saying to them, be fruitful and multiply, fill the earth ... God saw all that he had made, and indeed it was very good.*

This day is 'very good', reminding us of the 'Thank God it's Friday!' feeling. The third and sixth Days get extra blessings: the numbers of harmony, the trine and sextile aspects are stress-free. Also on the 6th day, *to you I give all the seed-bearing plants everywhere on the surface of the earth, and all the trees with seed – bearing fruit, this will be your food"* Both the Mars and Venus Days of Creation are concerned with seeds and fertility.

<u>The Saturnine seventh Day</u> is connected with Time and Memory, when the Elohim look back at the earlier steps of Creation:

> *And on the seventh day God ended his work which he had made; and he rested on the seventh day from all the work which he had made. And god blessed the seventh day, and sanctified it.*

This sequence has the *same seven archetypes* as are expressed in the days of the week.

A song by the *Incredible String Band* called 'Creation' echoes this sequence in a remarkable manner, and also links up to our metallic theme (enjoy listening to it on Youtube):

The first day was golden
And she colored the sun
And she named it Hyperion
And she made it a day of light and healing

The second was silver
And she colored the moon
And she named it Phoebe

And she made a day of enchantment and the living waters

And the third was many-coloured ♂
And she colored the earth
And she made a day of joy
With the scarlet strength of seed.

In the fourth black and white were mingled into quicksilver

And she colored Mercury ☿
And she made a day of wisdom
And the signs that are placed in the firmament

The fifth was bright blue ♃
And she envisaged Jupiter
And she made a day of awe and circles, circles
And she sent it to guide the blood of the universe

The sixth was burning with icy, green flames that glowed white

And of her beauty she made Venus ♀
And she made a day of love
Whereby all beings are united

The seventh was rich purple of the mollusks ♄
And she colored Chronos
And she made a day of idleness and repose
Whereon all beings cease from struggle.

Creation, The Incredible String Band, 1969

I met Robin Williamson who wrote this song, while he was teaching a bardic singing course at Emerson College (where I once studied) in the 1990s. I questioned him about it. He struggled to recall it ('Hyperion...'), but in vain. It was like some other vanished lifetime of his, it had gone...

3. GOLD, SILVER & MERCURY

The Ladder of Perfection

We have looked at some sevenfold sequences, which interlink metals, planets and days of the week, rather mystically. Next we come to what alchemists reckoned was their *scale of perfection*, from the base metals right up to the most perfect metal, gold.

− 1.50	Gold	☉
− 0.80	Silver	☽
− 0.79	Mercury	☿
− 0.3	Copper	♀
+ 0.04	Iron	♂
+ 0.14	Tin	♃
+ 0.13	Lead	♄

The glyphs, on the right-hand side, have been applied to the seven metals for *at least* as long as they were to the planets, if not longer (Chapter 8). To the left are electrode potential values (in volts), which indicate how chemically reactive they are, how readily they will dissolve in a dilute acid.[31] The more noble or *perfect* a metal was for the alchemist, the less *reactive* it is for the chemist.

Alchemists believed they could help metals to 'mature' and that meant moving up this scale of perfection. Mother Nature did this underground, or so they believed, she was continually making gold and silver, but slowly and over long periods of time. The alchemist would just speed this up a bit.

Which of these do you like wearing? I quite like copper, but I prefer silver. If your kid likes wearing iron décor you are probably in trouble, he or she may be quite disturbed and/or disturbing, and may get into street fights. In Britain, places of worship have the base metals iron and lead on their outside, and are in consequence quite

---

[31] 'ChemiWiki Standard reduction potentials by element', taking the highest-valence redox potentials, for values here used. These are also the values Frank McGillian gave in *The Opening Eye* (1980).

depressing. I remember visiting the city of Sophia in Bulgaria, where the churches – of Eastern Orthodox faith – had lots of gold and copper on their outside. It made a happy mood in the city.

The Bible gives us six of the seven metals, in the correct order. The Book of *Numbers*, Ch. 31:

> **22** Only the gold, and the silver, the brass, the iron, the tin, and the lead,
> **23** Every thing that may abide the fire, ye shall make it go through the fire, and it shall be clean: nevertheless it shall be purified with the water of separation: and all that abideth not the fire ye shall make go through the water.
> **24** And ye shall wash your clothes on the seventh day, and ye shall be clean, and afterward ye shall come into the camp.

One doubts whether that text can be earlier than the 1st century BC: to have the seven days of the week plus that metallic sequence. Some may believe that this was written in 1400 BC by Moses, or in the 6th century BCE or whatever and one appreciates people want and need to believe these Hebrew texts are enormously old. Early alchemical beliefs developed in Alexandria, as did the seven-day week, which is where the original Greek texts of the Old Testament appeared in the first and second centuries BCE. The earliest dateable day-of-week was in 30 BCE, in Alexandria, since when the seven days of the week have been cycling without a break.

The 'Seven steps to heaven' derived originally from the Migthraic religion which we look at in Chapter Eight, in the above order. The metallic *scale of perfection* pertained originally to a ladder or *stairway*

to heaven. Here is a fine diagram but alas its steps are in the wrong order! Still, at least it has gold at the top and lead at the bottom.

Gold and the Sun

There's a lady who's sure
All that glitters is gold
And she's buying a stairway to heaven

Led Zeppelin

Traditionally the noblest of the metals, gold expresses the splendor and radiance of the Sun. As the only metal which never tarnishes, it will resist the fiercest fire. Its sun-like nature is evident, for it needs to glitter in the sun to express itself, and has a unique relation to light and color. The metal can be beaten out so thinly that it has hardly any solidity left, when it appears as gold by reflected light but green by transmitted light. Colloidal gold solutions, in dilutions of parts per 100 million, produce a wide variety of colors. From metallic gold one can obtain, so to speak, any color under the sun: 'In gold we see the brilliance of the sun, but other rich colors are also seen in its colloidal solutions, ranging from greenish-blue, through reddish, violet-blue to pure rose – from the gold of a noonday sun to the radiant colours of sunset.'[32]

The sun manifests the colour of gold at sunrise and at sunset. The latin word for gold, aurum (thus, the chemical symbol Au), derives from the Greek word Aurora - the golden goddess of the dawn. Rudolf Steiner gave 'AU' as the Sun-sound, so try intoning it. The word 'aura' comes from the same root, indicating the idea of radiance as associated with this metal.

For the lovely hues of glass colored with gold - about ten parts per million of gold gives its pink hue.[33] Normally people will never recognise this color, will never guess that gold is producing it.

---

[32] Alison Davidson, *Metal Power, the soul life of the planets* 1991, p10.
[33] Guy Desbiolles uses this recipe (see Appendix): Prepare a 0.001 M gold chloride solution, and a 1% trisodiumcitrate solution. Take 5 ml of the first and add between 0.1 and 1 ml of the second one and heat the mixture in hot water 85°C for 2 minutes. The mixture turns first blue-grey, reaching its end color after a few minutes. See also, Youtube 'Ruby Red Colloidal Gold.'

Like sunlight through air, so is gold diffused through Earth's crust: 'Gold is a remarkable substance. A description of its physical properties can leave one in awe, even disbelief. Gold is present everywhere on the Earth - in the seas, in the highest strata of the atmosphere and in the earth itself on every continent. It exists as the finest dust and dense nuggets. There are however no veins of gold as there are of other metals. Gold is to be found finely distributed, combined with silver, mercury, copper and antimony.[34]

The gold mines in South Africa descend thousands of metres, to mine gold present in maybe less than one part per hundred thousand of the ore - only to be reburied in bank vaults! The largest deposits of gold are found in Africa. In this continent, whose geography shows so many different sun-influences, and whose music expresses so powerfully the throbbing pulse of the heart, the greatest amounts of the sun-metal have condensed.

Gold prices tend to rise in times of turbulence and uncertainty. The price of gold has a seasonal trend (Appendix 2), peaking just after midwinter. When sunlight is weakest, people show a greater desire for gold, the sun-metal.

Gold is a metal on a journey, shown by its number given in 'carats,' which goes up to twenty-four, for absolute purity. A gold ring may be 18 carats, and thinly-beaten gold which needs to be soft could be twenty-two carats. The carat-number indicates how long the gold has been in the furnace, how intensely purified it has been to free it from baser metals. The Sun moves across the sky every twenty-four hours, and around the year in twelve months, so this solar number defines the quality of gold. The weight of gold is measured in Troy, with one Troy ounce of gold equivalent to 480 (24 x 20) grains of wheat. The golden grains of wheat, sun-ripened, are fixed in an equivalence to the solar metal, indicating a healthy basis for currency and wealth.[35]

Most of us never get to experience the weight of gold, it is far denser than lead or mercury. The character 'Auric Goldfinger' in the James Bond movie declares that gold is attractive due to 'its brilliance, its colour, its divine heaviness.'

The well-being of a culture should be measured by the proportion of gold which it keeps above-ground, to glitter in the sunlight and

---

[34] Christopher Budd, *Of Wheat and Gold*, 1988, p.47.
[35] Ibid, p.50.

adorn the beauty of womankind, its sacred temples and places of magnificence - as compared with that buried in vaults and hidden away underground. The former indicates a commitment to communal happiness, and in fact solar glory, while the latter embodies private greed, lust for lucre, and hidden control.

Until very recent times gold was used as a heart remedy, this being the organ associated with the sun. Homeopathic doctors still use it in this manner in high dilutions and regard it as a remedy for depressive or suicidal conditions: a 'total eclipse of the heart'. Its distribution within the human organism reaches its highest concentration in the region of the heart. Gold is used by doctors to diagnose heart problems. As the highest concentrations of gold in the human body occur around the heart, a radio-isotope of gold has been developed (the Au-195 isotope), which can give an image of the blood-containing structures within the heart, a process called 'heart-imaging'. Gold gives a heart image! In Britain this technology has been developed in St. Bartholomew's hospital, London. One expert described the gold used in this way as 'a very convenient medium for rapid assessment of changes in cardiac function.'[36]

Economically, gold functions as a kind of heart-centre which maintains and guarantees a circulation of paper money. The pulse of the economy is taken by noting the value of gold. In ingots of great density it is stored underground, far from the sunlight; where it acts as it has done throughout history, in a somewhat magical manner, as that which is most to be desired - again, a heart quality held by no other element. After all, would you want a wedding-ring of platinum?

Astronomers are discerning how the Sun functions as the heart-centre of our solar system (though they don't see it in these terms). It has a heart-beat over its twenty-two year cycle and thereby circulates material around the solar system. It's the fiery heart-centre of the macrocosm, and gold is the Sun-metal. Feel the fiercely burning solar corona around, your heart of fire. Gold works as a heart-medicine.

Cows have the highest concentration of gold in their horns. The horns are the one part of a cow that points upwards, which give to the cow its dignity. Gold has no biochemical purpose, because it is

---

[36] Elliot et al., *Physics in Medicine and Biology*, 1983, Vol. 28, pp.139, 147; Dymond et al., Journal of the American College of Cardiology, 1983, Vol. 2, pp.85-92. (Source: Dennis Elwell, *Loom of Creation*)

chemically inert. But, let's go a little deeper and envisage the Egyptian image of the Sun between two horns of a cow. I have a dream, of a laboratory, where the apparatus is simple enough to bring delight to a child. It would have two Perspex models, of a human being and a cow, showing their varying gold-concentration: reaching its highest level around the heart for man, and in the horns for a cow.

The Sun's position in a birth-horoscope is said to express one's true being. To get a focus on this, you might want to consider why people spend - or rather used to spend - more to have a pen with a gold nib. This isn't just because it lasts longer than a steel-nib, but because of something not easy to express, that handwriting with a gold-nib pen better expresses one's 'personality' or inner being than does a steel-nib pen.

Medicinal Colloidal gold is becoming more widely used as a medicine – a very traditional alchemical concept. It is claimed to work in a quite subtle way as a heart-remedy. Some find that their will-power is enhanced upon taking it, in terms of being able to focus on what one wants to achieve. Especially in America there has been a tradition that colloidal gold is given for 'dypsomania' or craving for alcohol. Drug-addicts are said to experience a loss of appetite for their drugs after taking the solution for a few days. Here is a web-testimony: "I have found great benefit personally in the emotional area. My wife will tell you I am a much easier person to live with ... My brother who is four years older was in such a bad state the doctors had him on Zoloft, the antidepressant that is like Prozac. He started on colloidal gold almost two years ago and is living a happy life now with no known side-effects." It is said that the body's warmth-mechanism may be positively affected by gold, particularly in cases of hot flushes, chills and night-sweats.

One would like to hear more discussion of these effects, as may deepen our insight into how traditional heart-qualities are associated with gold, the radiant Sun-metal. Colloidal gold solutions here show the different hues, and gold foil less than a micron thick is here shown glowing green by transmitted light. This is what you should have been shown in your school chemistry lesson.

Silver and the Moon

'Celestial Diane, goddess argentine
Shakespeare, Cymbeline

The pure silvery Moon was associated with the chaste Moon goddesses, Artemis, 'the Huntress with the Silver Bow', and Diana, whose images were cast from silver. The silversmiths of Ephesus who made such images are referred to in the New Testament.

Today, in the delicate chemistry of silver we may trace its Moon-nature. It is a metal which requires darkness for its reactions. A photographer needs darkness in his studio to work with this metal. Special bottles and pipettes made of dark glass are used for solutions of silver, and its salts are quickly spoilt by exposure to the light of day.

Silver and gold are the two metals which show an intimate connection with light in their chemistry, although in opposite ways. The Sun produces the different colours of day, whereas the Moon shining only by reflected light gives the black, white and grey tones of a moonlit scene. Gold itself produces the different colours, one feels its outgoing radiance, whereas silver receives light images passively, it is precipitated from solution by light. The silver images of photography are only in black white and grey, and for colour film salts other than silver must be used.

Astrologers associate the Moon with the faculty of imagination, of fantasy, as for example in imaginative writers or dreamy poets. The same property is seen in the way silver is able to create images. In photography it creates a memory-image of the past, in mirrors it gives an image of what is in present time before it. Today, most mirrors are made by coating glass with silver. When looking at a mirror we never feel we are looking at a sheet of silver. There is a certain receptiveness and passivity here, and similarly when looking at a photograph it never occurs to us that we are really looking at the differential precipitation of colloidal silver. We are not aware at all of the metal but only of the image it provides.

Silver is used by the cinema industry to form its 'images of the silver screen'. Silver has always been the staple metal used for making films, in colour as well as in black and white, and the film industry is a major drain on the world's silver reserves.

From an astrological viewpoint, one can say that the dreams and fantasies which the cinema manufactures are somewhat lunar in nature, because the Moon is associated with dreams and the imagination. By its delicate and receptive Moon nature, the metal silver, in celluloid, will faithfully record light images.

The metal chromatography techniques developed by Kolisko are another example of the image-forming powers of silver. Here the varying images built up by the precipitation of colloidal silver are produced not by light but by the changing conditions of the cosmos itself. Properly used, this technique is an empirical method of investigating the correspondences here described. Silver's Moon-quality of receptiveness here manifests remarkably.

Figure: Electrolytically refined silver (Wiki)

A nice point was made by the reviewer of Agnes Fyfe's work *Die Signatur der Venus im Pflanzenreich* (The signature of Venus in the plant-realm).[37] This follows on from Fyfe's previous work, *The Signature of Mercury in the Plant-Realm*. The reviewer pointed out that Kolisko's work used 'above the Sun' metals, iron, tin, and lead, whereas Fyfe's work with plant sap uses the 'below the Sun' metals,

---
[37] The Astrological Association Jnl., Spring 1980, p.50.

copper and mercury. So all seven of the metals have now been used for recording chromatographically cosmic events of their associated planets. In both cases silver is normally used for manifesting the images, although in the latter case gold can be used if primarily colour rather than form is desired.

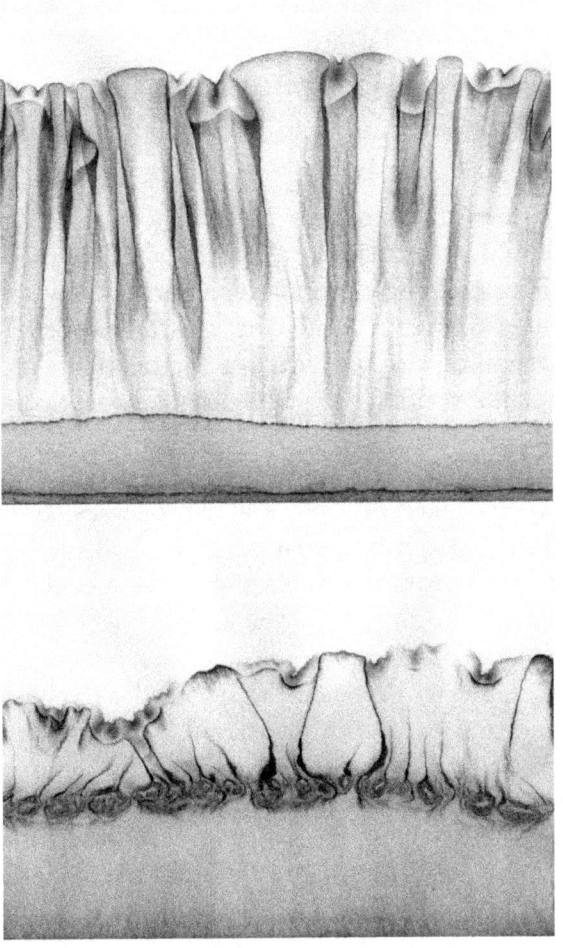

Figure: 'Steigbild' filterpaper pictures using plant sap and silver nitrate

In these two images, prepared by Guy Desbiolles in Switzerland, the receptive Moon-metal silver is giving expression to the 'formative forces' in plant sap. Sap from a stinging-nettle is extracted in a pestle and mortar, and diluted with water. Then it is risen up a filterpaper, and it dries. Then the next day 1% silver nitrate is risen up through this dried plant sap. That is the basic method, developed by Kolisko. The top image was done at a Full Moon, the next at new Moon. I will certainly endorse the basic idea which Guy Desbiolles is here showing, of a different quality-of-being at Full moon compared to New, in the plant realm: yes!

Any alchemist could tell you that. We may here meditate on what Kolisko wrote about the metal silver:

> ...silver is a metal which has in itself a hidden power of formative force which we do not find in any of the other metals in the same strength.[38]

Chapter ten discusses the experiments designed by Kolisko, whereby celestial events could be recorded on filterpaper.

Silver is a mirror-creating element: a solution of silver in a test-tube readily precipitates a mirror onto the glass, this being the school chemistry test for silver in solution. As a metal it has the highest electrical and thermal conductivity of them all, as well as being the best reflector of visible light known.

Most of the world's silver occurs dissolved in the oceans, reminding us of the Moon's connection with water-processes. Silver iodide is used to make rain, by sprinkling it as a fine dust onto rainclouds, which leads to condensation. Shakespeare called the Moon 'Pale governess of floods', and rainfall as well as the tides has been shown to vary with the lunar cycle.

In the 1950s, ionic silver began to be used as a bacteriocide for water purifying systems, in the form of a precipitate on carbon granules. A U.S. Navy study, using ships passing through contaminated waters, found that a silver concentration of ten parts per billion made the water safe for drinking (homeopathically, a D8 concentration), and this method is nowadays used by shipping companies. Good domestic water-purifying systems nowadays contain,

---

[38] E. and L. Kolisko, *Silver and the Human Organism,* 1978, p.9 (she co-autored this with her husband).

as well as an ion-exchange system, a silver tube which acts as a bacteriocide.

It has long been known that water carried in silver flagons stays fresh. Settlers moving across the American West would purify a container of water by leaving a silver dollar in it overnight. At the John Hopkins University of Maryland, researchers kept a community swimming pool clean just with a carbon-silver purifier. A report concluded, 'During the time the silver-carbon filter was in operation, there were no cases of ear infections or eye irritations. Bathers and, in particular, swim teams enjoyed the clean, crystal clear silver-treated water without the usual disinfectants that sting, irritate the eyes, bleach swimsuits and affect hair colour'.[39] Here we see silver's bacteriocide action, its action as the Moon-metal upon water, maintaining its quality. But, silver's Moon-quality of purity can be appreciated in other ways, as in the special sound of silver bells.

From such considerations we see how the following adjectives apply to silver:

Reflective, image-forming (imaginative), receptive, impressionable, sensitive, pure

Are these lunar traits? I think they are. Compare them with a list of traits which the Gauquelins obtained in their attempt to define a 'lunar personality':

Doux, impressionable, nonchalant, parle bien, reveur, sensible, spirituel, subtil, sportif (pas)[140]

-a modal personality which they found most pronounced in imaginative writers, poets and dramatists.

Modern Uses Nowadays, the main growth in silver markets comes from its use in jewellery and ornament – mainly in India. The 1990s have seen tremendous growth in this Indian market, much in the form of heavy-weight investment jewellery – bangles, ankle-rings and necklaces. Virtually every Hindu woman wears an ankle chain, which is nearly always silver. How appropriate that the Moon-metal should

---

[39] *A Brief History of Silver in Water Treatment*, p. 10, John D. Collins, Ionics, John Hopkins University.
[40] 'gentle, impressionable, nonchalant, well-spoken, a dreamer, receptive, spiritual, subtle, not the sporting type', M. Gauquelin, *La Cosmopsychologie*, Paris 1974, p.155.

be used in these feminine and decorative contexts. Many domestic and decorative utensils, often given at the time of marriage, are silver, as likewise are those used in devotional ceremonies. Muslims use much less silver because of strictures imposed by the Koran, which seems odd considering the lunar symbolism inherent in Islam.

Photography is the main use for silver, despite competition from digital cameras. The firm Britannia Refined Metals in Kent extracts around 500 tonnes of silver from crude lead per annum, using lead shipped over from Australia. It refines the silver to 99.9% purity and then sells it to London bullion markets. I once visited the London Metal Exchange, and watched how the trading of silver was shown by a crescent-Moon glyph! The LME is the world's largest centre of trading for non-ferrous metals. Metal-dealers have always used the traditional alchemical glyphs.

The healing properties of silver appear as rather maternal and protective as befits its lunar essence. Colloidal silver is regarded as safe to use during pregnancy and lactation. Here's a website-testimony of a cure, by 'Jeana' who was ill and sore with mastitis - "I was very, very ill and the antibiotics were not helping at all," and then her father suggested she try some colloidal silver. "Being a sceptic, I drank the glass of silver very slowly, trying to taste any strange aftertaste etc. ... By the next morning, my mastitis was completely gone. My breast was no longer red or swollen. My baby had not been nursing well at all during this bout of mastitis; I had to use hot compression to get even one drop of milk from my right breast. But by the next morning my milk was free flowing, and I felt great." Phew, if that isn't lunar symbolism tell me what is!

Silver has a dynamic way of healing injured and damaged tissue. This was pioneered by Dr Robert Becker in *The Body Electric* of 1973. He found that by using silver electrodes he could stimulate bone-forming cells and stimulate healing of the skin and soft tissue, as described in a recent review by Mr Best: "Partly as a result of Becker's work silver has been used in bone healing for many years now and is incorporated into bandages to speed up healing ... a recent US study reported that silver catheters can prevent urinary tract infections much better then uncoated ones."[412]

There is a resurgence of interest in colloidal silver for combatting

---

[41] *Caduceus,* Autumn 2001, Report on Colloidal Silver pp.31-35

colds and viruses, and it could well be that there is no more useful bottle to have available in the family medicine cupboard. To quote Best again, "Ongoing research may eventually restore silver to its once-accepted status as probably the most versatile and effective natural agent against bacteria, fungi and, recently, viruses available – with the hugely important bonus that the latter finds it almost impossible to develop any resistance to it." Also, colloidal silver is being used in skincare creams for its antiseptic properties, being effective on oily skin which is prone to spots and itches. "Colloidal silver is Nature's antibiotic and has an antimicrobial effect," a herbalist at the London *Fermacia* clinic remarked, adding that it purified the skin.

Doctors are likely to continue opposing these gentle, healing powers of silver, because their theories can't account for it. As Luna (the Moon) tends to elude the categories of rational explanation, remaining enigmatic in a manner that baffles astronomers, so her shining metal silver may do likewise.

Quicksilver and Mercury

'A mind like quicksilver'-how well this image applies to mental processes! It is hardly surprising that astrologers should associate the planet Mercury with mental agility: the shining globules of this liquid metal form and reform so quickly, as fast as thinking. The metal mercury is the one element that one normally sees in the three states of matter - as the fluorescent lamp overhead in the classroom, as the liquid in the thermometer and as calamine the skin lotion; as Hermes was the one deity who could come and go through the three worlds.

Alas, the nimble quicksilver intelligence can end up as the 'mad hatter,' whose mind is a-jumping all over the place - remembered in Alice's immortal tea-party. This was a condition to which hatters were prone in Victorian times, due to using mercury metal to give a shine to top hats.

As Hermes was the messenger of the gods, so mercurial types make good link people. Likewise the metal mercury amalgamates: different metals can be brought together by dissolving them in mercury, it is a solvent for metals. The term 'amalgamate' is also used in commerce: different firms amalgamate together. This is a mercury-process, and Hermes was traditionally the god of commerce.

The most characteristic chemical trait of mercury is association. It links itself up in the most unexpected ways. 'The tendency to form complex compounds is very marked in the case of mercury.'[42] It combines with nitrogen and carbon compounds which metals normally won't touch, as well as forming the usual metal salts, and forms complicated 'organometallic' mercury compounds, which catalyse the synthesis of a range of pharmaceutical and other organic, man-made products. It forms explosives (e.g. mercury iodide) which detonate at a mere touch. In amalgamating other metals together, it performs this interlinking function.

The Indian word for alchemy was 'Rasayana' which means 'the way of mercury.' The earliest alchemical texts in the West date from the first century AD, and this is also when the first texts for obtaining mercury from its ore cinnabar appear. Pliny the Roman naturalist gave such a recipe. Heating of the red ore cinnabar causes it to sweat globules of the shining metal; then, careful heating the mercury again yield a red ore (although this is the oxide, no longer the sulphide). This was the classic recipe whereby alchemists impressed their clients, and was the first inkling of a chemical reaction. Mercury's changeable nature seemed to manifest mysteries of matter. Hermes in his Egyptian form as 'thrice-greatest' was the patron of alchemy, in which mercury had the central role. Alchemists who reckoned they could make gold would usually start off with mercury (which is, as chance would have it, next to gold in the Periodic table)

The orbit of this fastest-moving of planets was an enigma for a century. The plane of Mercury's orbit kept 'precessing' or shifting about in a way that defied explanation, and Newton's theory could not account for it. Mercury resisted this materialistic world conception, and it was only explained in the 20th century by the Theory of Relativity. Likewise, the metal mercury resists the solid state. It is the secret, the mystery of quicksilver, that a metal of such enormous density can yet remain liquid. It is not difficult to see why the alchemists credited mercury with a very special inner mobility and vitality.

---

[42] J.R.Partington, *A Textbook of Inorganic Chemistry* 1960 p.795.

The commonest daily use for mercury sees it in constant motion - the thermometer. Hermes was traditionally the god of medicine, and Mercury was for long given an important role in medical practice. It was for centuries the staple remedy for syphilis, and even today it is still used for skin ointments-calomel-and the sublimate is used as a disinfectant. Mercury amalgams are used in dentistry, and mercurial aids such as the thermometer and blood pressure apparatus aid the doctor. Thus the different aspects of the Mercury-nature are expressed both by the metal and by the planet in the sly, in accordance with the Hermetic maxim, 'as above, so below.'

Mercury is always on the move, and nowadays it is coming out from circulation: from batteries, from tooth fillings, from gold amalgamation processes, etc, so that Euro- experts have a problem what to do with it. Thousands of tons of it might be placed carefully down one or two of the mines whence it was obtained! As Mercury is removed from large-scale use, we may be sure that other subtle properties of this mysterious and elusive element will turn up in due course.

4. Copper, Iron, Tin & Lead

Copper and Venus

He learned chemistry, that starry science

Moffet's biography of Sir Philip Sydney[1]

On average, women have about 20% higher copper in their blood serum than men and for iron it is the other way round, with men having a one-third higher iron level than women in their blood. The deep significance of this fact is entirely ignored by modern medicine. Iron and copper levels are sex-linked in exactly the way expected from the gender symbolism of their planets. The level of copper in human blood is critical, being around one part per million by weight, and normally it remains fairly steady around this value.

Copper in women's blood serum has a monthly cycle in tune with their menstrual period, peaking a week or so before the period arrives. This is because their serum copper exists chiefly as the protein, 'ceruloplasmin', whose metabolism is closely linked to the female sex hormone oestrogen. The Pill works by emulating conditions of pregnancy where oestrogen is high, and this has a drastic effect upon serum copper levels. During pregnancy, copper serum in the mother climbs up to <u>double</u> its normal level, reaching 1.9 parts per million. Conversely, iron in foetal blood also increases as the time of birth approaches, so a copper-iron polarity develops between mother and child. Insomnia, depression and changeable moods towards the end of pregnancy have been related to the raised copper levels. A woman taking the Pill has blocked off her monthly rhythm of serum copper, and instead retains a permanently high level corresponding to the

---

[1] Thomas Moffet's *View of a Life and Death of a Sydney*, quoted in Charles Nicholl, *The Chemical Theatre* 1980, p.15.

ninth month of pregnancy. Evidence suggests that copper has a dynamic role in the reproductive process, rather than just being a by-product of the raised oestrogen.

In the early 1970s it was discovered that coil contraceptives using copper were much more successful than previous coil designs. The 'copper-7' coil became the most popular design and was marketed world-wide, used chiefly by women who have already had one child. Despite intensive research however, no-one had any idea as to the mechanism whereby copper in the coil helped prevent conception. Copper ions have a biological action on the inside of the uterus, preventing implantation of the fertilised ovum. Its modus operandi is thus quite unconnected with that of the Pill, where overall blood serum levels are raised. The sole connection is that in both situations a striking Venus-quality is shown by copper's behaviour.

Figure: malachite ore

Having compared copper and iron in the blood, let's compare them in other aspects - as their two planets are nearest to us, one within Earth's orbit and the other outside it. Pure copper is a metal of reddish-pink hue, and has a warm, beneficial glow which contrasts with the cold glint of steel. With something made out of iron one may feel 'how strong' or 'how useful', whereas with something made out of copper, the first impression is more aesthetic. Whether it is a copper bowl, a trumpet, or a green-domed copper roof, it is the visual appearance rather than the utility of the metal which first strikes one. It is such a soft and pliable metal that it needs to be alloyed with other metals, into brass or bronze, before it can be used for a structural purpose.

In an exhibition of mineral ores those of copper first attract the attention, providing a joy to the eye as do those of no other metal. Look at the delicate green-blue hues of malachite or azurite- how different from the massive, solid forms of the iron ores, pyrites or haematite! The pyrite crystals form perfect cubes, expressing Martial power and strength. A contrast to this is the copper ore malachite, often cut and polished for decoration, to disclose its swirling patterns and sea-green hues. The names of the ores of copper point to gentle Venus qualities: azurite, malachite, turquoise, chalcopyrite and peacock ore.

A room in which iron or steel predominates has the atmosphere of an office or a factory. It demands a mood of efficiency from us. A room in which copper predominates, in contrast, has a warm, homely atmosphere, in which we can relax. This is a key concept to the English pub. Americans don't understand this, and have drinking-bars where the cold glint of steel is evident, as promotes their violent and restless society. The high resonance of copper makes it suitable for a

Maxfield Parrish: his poster-style American art glowed with a tranquil and paradisal copper-Venus light.

wide variety of musical instruments-in the strings of a string instrument, in the brass section, in percussive instruments, and so forth. Traditionally astrology associates the arts of music with Venus.

No-one has better appreciated the glowing hues of copper than the American artist Maxfield Parrish. His natal chart (25 July 1870) had strong Venus-aspects (it was conjunct Mars and Moon and in opposition to Saturn). One of his pictures is here shown.

To trace the connection of copper with Venus we have to go back to a distant mythological era: back, in fact, to a Mediterranean isle, once ruled by a love-goddess - the island of Cyprus. This island was regarded as the domain of Venus-Aphrodite. Aphrodite was referred to as the 'Cyprian goddess'. In Botticelli's picture, The Birth of Venus, she is depicted as being born from the sea on to the shores of Cyprus. It is from the name of this island, Cyprus, that the word copper derives. The word copper comes from the Latin word cuprum and this derives from the Greek work Kyprus. Cyprus was in antiquity the principal source of copper, and so the metal was named after it. Venus was felt by antiquity to dwell just where such large amounts of copper had condensed. Venus was credited with a sea origin, and copper reminds us of this connection with the water element. All copper salts are sea-coloured, blue or green. All the ores and all the salts of copper are hydrated, water containing. Nearly all copper salts are highly soluble in water. The iridescent hues of a peacock's tail (see picture) derive from green-blue copper complexes.

In various sea creatures the breathing process is by means of copper, not iron. They do not need the fiery Mars-energy, but have a more tranquil mode of being. A simpler, copper-containing molecule is used instead of the iron-molecule haemoglobin. The conch shell in Botticelli's picture, always traditionally associated with Venus, comes from such a creature, one which respires by means of a copper-process. The octopus and the scorpion both respire using a copper-molecule in place of the iron-based haemoglobin.

That same polarity functions in an inorganic realm in the principle of the dynamo, where the relative motion of iron and copper generates electricity. Iron creates the magnetic field and copper wires carry away the current generated. The energy powering our civilisation derives from a pulsating Mars-Venus interaction, making alternating current. There was a Mars-Venus conjunction in the sky on the day when Michael Faraday discovered the dynamo principle (17 October, 1831).

As Mars and Venus in mythology were closely related, so are they found bonded together in the depths of the earth: the principle copper ore is in copper-iron pyrites, in which copper occurs together with iron. The darker threads of iron run through this sea-green ore of copper.

Modern uses of the red metal range from computer microchips to solar power cells, and it remains a key material for telecommunications, even though optical fibres are now preferred for trunk lines. A mobile phone has several grams of copper in it. There isn't a great deal of it left to mine - another about thirty years' worth, maybe - which has caused it to become a highly-recycled metal. Architects appreciate its pliability and visual appeal. Copper's lovely turquoise patina normaly takes a few decades to mature, from exposure to the elements, but modern techniques can accelerate this process into a mere couple of months. London's skyline has some fine copper roofing, e.g. on the Planetarium, Old Bailey and Royal Festival Hall.

Beauty creams use copper powder, notably the 'Dr Haushka' range and Weleda's Copper ointment: "Copper has a vital role to play in skin repair because of its ability to stimulate the growth of collagen and elastin... products containing copper tend to have good anti-inflammatory effects on the skin." How pleasant to hear of the Venus-metal's cosmetic use. After all, it is melanin, the copper-based skin pigment, which gives the bronze hue so vital for beauty's image - not to mention brown hair colour, also due to melanin.

See www.astrology-world.com/venus.html for the special harmonies of the Venus-orbit. Or, better still, get my book *Venus the Path of Beauty*.

Iron and Mars

He who knows what iron is, knows the attributes of Mars.
He who knows Mars, knows the qualities of iron.
Paracelsus

Red storms rage across Mars. The soil of the 'red planet' is high in iron, and its dust, swirling up into the atmosphere, causes giant storms that last for weeks. They blot out all its surface features, even the huge mountains. This can be quite a problem for craft trying to land there. One feels that these storms well express the Mars-quality of *anger.* Mars has around fifteen percent of iron in its surface soil, thrice its average level here on Earth. There must once have been lots of oxygen around, as all the iron is in the red, highly-oxidised (ferric) form.

Mars has two rocky little moons, and one of them is destined to crash in futurity, doomed to disintegrate, being too close to its parent planet. By meditating on these things one can experience the being of Mars. The archetype of Mars is fully expressed both in the red planet in the sky and in the metal iron under the ground. Thus, modern space discoveries have deepened our understanding of the primary Hermetic principle, as quoted above by Paracelsus.

Of the seven metals, iron is the 'earthy' one, having a stronger connection with the Earth than do the others – for a start, it's the only one that aligns with the Earth's magnetic field, in a compass. It's present in far larger quantities in the earth's crust than the other 'classical' metals. The others – lead, tin, gold, copper, mercury and silver – total no more than 0.01%, or one part in ten thousand of the Earth's crust. They are in a sense little more than 'visitors to the Earth' (the phrase is Dr Steiner's), although we use them so much nowadays that we forget their scarcity. Iron makes up about 5% of the crust, being the only one which has built itself solidly into the substance of the earth.

Reddish-looking soil means that iron is present, and, for the same reason, Mars is red. So the symbolism of Iron-Mars is direct and obvious, with nothing subtle about it. Mars has always been associated with blood and war because of this symbolism. But, let's not forget that that symbolism is also physiological fact: the blood is red because of the iron in it! The main ore of iron is pyrites, 'fool's gold.' Have some of its marvellous cubic-crystal structures on your mantelpiece! The other common ore is haematite, which has a quite different bulbous structure, and a dark reddish hue. 'Haem' means blood and 'pyr' means fire – blood and fire! These are indeed the Mars-attributes.

Of the seven metals, iron is the one that burns. A falling star is burning iron. A meteorite burns brightly as it falls through the atmosphere. Fireworks use the burning sparkle of iron filings. Some steel wool can be ignited, then plunge it into a jar of oxygen, when it will glow fiercely. Thereby one experiences the fiery energy of Mars. ... One is reminded of Vulcan and Ares by such a demonstration, the two Mars-archetypes of antiquity. One was married to Aphrodite the Goddess of Love while the other just had an affair with her. Vulcan or Haephastos was the Smith, who forged the armour and instruments of war.

The fiercely-glowing iron is removed from the furnace and then hammered into shape. Haephastos was lame, symbolizing an affliction that could befall Smiths from arsenic-poisoning. As copper and iron are bound together in the Earth, with most copper ores bound up with iron, as copper and iron interact in the blood, copper helping the iron metabolism – so Mars and Venus were mutually attracted. One sees this in the principle of a dynamo, where an iron-copper interaction takes place to produce the throbbing pulse of electrical energy.

Within the human body the fire-energy of Mars is seen in the metabolic process. Iron has a key role in the combustion processes within the tissues of the body, whereby food is turned into energy. Blood becomes red as the iron-containing molecule haemoglobin carries oxygen throughout the body. Thereby foodstuffs can be metabolised, as fuel for the organism. This same iron-molecule then absorbs the product of combustion, carbon dioxide, so the blood becomes blue, and carries it back to the lungs.

In this breathing iron-process, which never ceases while life is present, there is the restless energy of Mars, as the blood rhythmically alternates from red to blue, from oxygen carrying to carbon dioxide carrying, moving away from the lungs then back again. As it does this, the unique iron-molecule haemoglobin continually changes its shape. Iron has the property of readily passing from one valency condition to the other, as connects iron with the rhythmic breathing process.

Copper, iron, tin and lead

Figure: An Iron-man warrior

One is 'anaemic' if one lacks iron in his blood; or, equally, it may refer to a lack of Mars-attributes in the character: strength, courage and will. Weleda make a 'meteoric iron' remedy as is supposed to help with giving such strength to the will, for persons who feel a bit anaemic, or as if they cannot cope with life's challenges. According to Mellie Uyldert, 'If they [anaemic people] eat nettles or spinach, their zest for life returns,' these plants being high in iron. Popeye the sailor is a modern Mars-archetype, who gains his strength from eating a tin of spinach.[2]

As the Smith forged iron, with his hammer and anvil, so the development of the blast furnace required control over fire-processes. Iron has been closely involved with human history, as its strength gave men the power to dominate nature: 'The true Iron Age, led mainly by the peoples of the West, arises as a creation of iron and steel. Iron is harnessed to purely material goals. The shackles by which man has enslaved nature are forged of iron.'[3] As you press down the pedal on the gas, an iron apparatus controls combustion. The Mars-metal iron unleashes masculine power. For the history of warfare, the development of hardened steel was pivotal, and is today the element from which all weapons are made.

An iron-and-coal girdle surrounds the Earth, through Wales, England, Germany, France, Russia and America. Today we live surrounded by iron and steel, and within this will find no cure for the

---

[2] But NB spinach may not be high in iron - this may be a 'Popeye'-myth.
[3] Wilhelm Pelikan, *The Secrets of Metals*, 1973, p.85

rising tide of violence.

We note the mystery of iron: comparing the black hue of your wrought-iron garden gate with the gleaming steel of the car bumper, can they really be the same element? A few percent of carbon turns one into the other, as reminds us of Iron's strong connection with the Earth. Carbon the Earth-element makes all the difference to iron, giving to it the awful strength of steel.

Nickel Cobalt, nickel and manganese are metals with iron-like properties. They are ferromagnetic, are used in steel alloys and in fireworks, and have comparable physical properties. Dr Rudolf Hauschka regarded them as being 'the brothers of iron'. Nickel alloys are vital for the tough, high-performance equipment in today's planes, trains, cars and boats, eg the turbines of a power-station. Most nickel produced goes into steel alloy, where it adds corrosion-resistance and strength. For example, Japan has a fast, magnetically-levitated train that zooms around Tokyo, and it utilises this alloy in various key roles. Nickel has an ability to absorb kinetic energy in the event of a collision and this contributes to the safety of passengers. These are all fine, Martial qualities.

The Chrysler building, a historic landmark in New York city, was constructed in 1930 using nickel-containing stainless steel, for its roof, spire and gargoyles, and today it is as good as new. In Devon, the Eden Project used a lightweight nickel-steel tubular alloy to make its magnificent bio-domes. In Spain, nickel-containing stainless steel is being used for parabolic solar collectors, which are generating electricity at a lower cost than any photovoltaic system. Modern tidal-power and wind-power systems depend upon this alloy to survive the elements. All nickel products are recycled after use - including the kitchen sink. It has an energetic new use in metal hydride batteries, which are quick to recharge and will deliver give high voltages: they power everything from mobile phones to laptop computers, from digital cameras to cable-free power tools. Electric cars with these batteries can travel a couple of hundred kilometres on a single charge. Nickel is found in certain iron-meteorites. Things have come a long way since the 'Nickels' or mountain-spirits of mediaeval times gave to this element their name.

Tin and Jupiter

If you were asked for the main use of tin, what would you say? Probably, its use for preserving foodstuffs. Traditionally Jupiter was the preserver, and a well-placed Jupiter in the chart is alleged to preserve youth. Ale used to be drunk from pewter mugs, a tin-based alloy. Theatres simulate the sound of thunder by shaking a sheet of tinfoil, which produces a roar like distant thunder (Jupiter-Zeus was the god of thunder, and astrologers associate Jupiter aspects with rain and thundery weather conditions).

The last Cornish tin mine closed in 1988, but of late steps have been taken to reopen a one, and it looks as if the oldest Cornish industry is now being resurrected. The Roman deity Jupiter was associated with both the Etruscan sky-god Tinia as well as the Greek Zeus. The supreme Etruscan sky-god "was known variously as Tin, Tini tinia or tinis. This supreme sky-god was depicted with lightning-bolts, a spear and a sceptre. Basically he as the complete prototype for Jupiter" (web comment). What with early trade-routes to Cornwall from the Mediterranean to obtain Cornish tin, one suspects there may be a semantic link here, but if so it seems destined alas to remain conjectural.

Bronze, an alloy of copper and tin, has been used to make sculptural objects as early as the seventh millennium BC. Pewter was introduced into Britain by the Romans in the 3rd century AD, being made from Cornish tin and Welsh lead. Pewter tableware was once commonplace, but gave way in time to porcelain and other materials; is now becoming popular again for decorative materials, but without lead (antimony is used instead). The worshipful company of Pewterers launched a Millenium pewter collection, extolling its traditional virtues – solidarity, lustre, practicality of use and handling, no tarnishing – which do sound quite Jovial.

Tin wine-capsules (wrapped around the necks of wine-bottles) now account for some 4,000 tons per annum of tin production, a growing market, and this seems Jovial enough. The Oscar-award trophies are made from tin, coated with gold. Use of tin solder is still growing in electronics, especially computer hardware, and laptop liquid crystal screens have a tin-oxide film for their panels.

There actually isn't a lot to say about tin and Jupiter. I once asked an old astrologer about this and he replied, why, how could it? The

Jovial qualities of optimism, expansiveness, balance, how could a metal express these?

Lead and Saturn

The toxic effects of lead have been known for a long time, saturnism being one of the first industrial diseases to be recognised.
    E.E.C. directive on lead in petrol, 14 December 1973.

The condition of 'Saturnism' was so named because its symptoms – headaches, fatigue, irritability and depression – seemed reminiscent of a 'saturnine' humour. From brewing cider in pewter vessels, to using lead paint on houses, and lead solder in water supplies, the addition of lead to petrol was just one more link in a long tradition. In the 1970s, scientists finally became able to measure reliably the level of lead in the normal-population bloodstream, at around one or two-tenths of a part per million (That's D7 in homeopathic terms). The idea that this could be impeding child intelligence and promoting hyperactivity seemed like science fiction.

I was then working in the government's Air Pollution Unit in London, measuring city air lead levels. It dawned on me that these symptoms had a 'Saturnine' character – as, equally, did the reason for putting the lead into petrol, which was to slow down the combustion-process and make it more regular. I participated in the great British lead debate which then erupted, and wrote a book on the subject, then finally apprehended that this debate had entirely revolved around the triple Saturn-Jupiter conjunction of 1980 (such triple events being very rare).

The issue then vanished from public debate, once the conjunction was over! (Appendix 1 describes this)

Lead stores in the bone tissue. If you could see only that part of a person where lead was, you'd see – a skeleton. Saturn-Chronos as 'old Father Time' was traditionally a skeletal figure. It takes about thirty years to flush out bone lead, so it's a fairly permanent affair. However one should not view this with 'leaden-eyed despair.' No other metal tends so much to form insoluble compounds. It is a heavy, dark, sluggish metal, and of the seven it is the slowest conductor of electricity and the least lustrous or resonant.

Lead exists as a boundary or limit of heaviness. To see this we need to look at the Periodic Table of elements, and there we notice that the elements above lead are radioactive, i.e. they are too heavy to remain stable. They decay, until they reach lead, then they stop. So various lead isotopes can be present in a rock, that are the end-products of radioactive decay. Broadly speaking, the more of these isotopes there are, the older the rock is likely to be. Scientists measure age by lead! They may take a ratio of uranium to lead, and from this proportion they will infer billions of years of age. This indicates Saturn's connection with Time, though whether such inferred eons of 'lead time' are real or illusory is another matter. When scientists looked at the moon rock, the most distinctive 'fingerprint' they found was that of the lead-isotopes, from which they inferred its enormous age.

The Saturn-archetype is associated with the notion of a boundary or limit, which reminds us of this lead-property. Also, Saturn holds an hourglass: use of lead in paint is now banned, but no other white paint can last through the years like a lead-based paint. But, lead paint is still used for painting the yellow lines along the kerbside (lead chromate). Only a lead paint is tough enough! Here the lead sternly prohibits one from parking, over certain limited hours, which seems Saturnine enough. The planet Saturn displays a strong vertical axis, with its magnetic and rotation axes coinciding within one degree, and at a right angle to these are the rings, exactly level and circular. A plumb-line holds the vertical, lead's earliest use (plumbum, lead Latin). Saturn's *gravitas* is sometimes required to hold one's balance in life.

Uranium, the first metal heavier than lead to be ascertained, was discovered in 1798, soon after Uranus, the first planet beyond Saturn to be discovered, and was named after it. One could say that a lead-Saturn limit was then passed. The disruption of accepted notions which ensued from investigating the properties of uranium hardly needs emphasizing. Today, nuclear power plants are shielded with lead, again reminding us of its connection with these extra-heavy elements.

Why does a car battery need to be so heavy, with so much lead in it? No other kind of battery would have the right time-properties, such as being able to deliver power over days without any change in voltage, unvarying; or be charged up over days then to give out that power in seconds. Here the inertia of lead is important; it is able to remain in the sulfuric acid for days without any reaction taking place, and only react when an electric current passes. Other batteries use two different metals, but in an accumulator the lead works alone, surrounded by acid.

As a heavy, dull, sluggish metal, it was traditionally the basest of the seven. Of the seven metals it is the slowest conductor of electricity and the least lustrous or resonant. No other metal tends so much to form insoluble compounds - an aspect of its heaviness. Use of this metal in paint is now banned, but still there is no other white paint that can last through the years as one that is lead-based. I was looking at an old sundial the other day, wondering why it was made out of lead... its more traditional uses include tombs and bullets.

In such mundane terms can one visualize the axiom 'As above, so below ' which alchemists of old attributed to the Emerald Tablet of Hermes. Imagination perceives what logic cannot easily deduce, that the function of these .metals shows their planetary essence. A California scientist Clair Patterson ascertained the age of the Earth by using lead isotope measurements. This shows the same Saturnine essence of lead in action as does its use in painting roadside lines. Both cases relate to time measurement -- one in hours, the other in billions of years!

Lead levels in human blood have been plummeting over the last two decades, right across Europe. The three 'p's piping, paint and petrol have all seen lead phased out and mean levels in blood are well below the one-tenth of a part per million (milligram per litre). So does that mean we're in the clear? Alas no - being so insoluble, lead takes a long time to leave the environment. Anyone committing a crime of violence is likely to have something like four times more lead and aluminium in their hair then a normal person - that's a staggeringly large differential. There is a very dark, negative side to lead poisoning and we need to understand this.

Lead has one of the highest recycling rates for all materials, with 90% recycling for lead in batteries for many European countries. Demand for lead continues to grow. The lead industry has been struggling to counteract negative public perception of this metal and an enormous body of legislation has developed to limit releases of lead into the environment. TV sets and computer monitors used to use cathode ray tubes and to protect viewers and operators from harmful X-rays, the glass used for the cathode ray tubes is one-quarter lead oxide.

By far the biggest use of lead worldwide is for the car battery, needed by every vehicle in the world – about three-quarters of total demand. This represents a huge increase in recent years. Experts are looking at ways of recharging these in minutes rather than hours. Lead-acid batteries are nowadays helping the two billion persons not connected to any power supply. They are maintenance-free, and can connect to renewable sources such as photovoltaic cells. In California, zero-emission electric buses can be recharged up to 50% in five minutes and such fast recharging is crucial for the bus's viability.

Levi's Periodic Table

An enchanting book on chemistry, so well written that it needs to be read quite slowly, is *The Periodic Table* by the late Primo Levi, translated into English in 1985. 'A narrative poem of magical quality' said the *New Scientist*; 'wonderfully pure, and beautifully translated' wrote Saul Bellow; while the *Washington Post* found it 'An extraordinary, nimble, fluent book from an extraordinary life, part autobiography, part fiction, but essentially something like a memoir of elemental matter.' But, were these reviewers able to put their finger on what moved them so? 'Flamboyant chemicals, sullen human beings; women living adventurously and organic chemicals that live timorously behind many locks, ready to bolt down the fire-escape when the analyst rings the front doorbell,' was the *New York Times'* account.[4] It would be a mistake to see this book as a mere collection of animated stories about chemicals ... though its author might, if pressed, have described it in such terms! No, it is the account of an industrial chemist's life who, looking back, sees himself as having grappled with primordial archetypes.

The chapter, 'Iron' is painted against the sombre background of a world at war. The author describes his friendship with a tough and wiry character called Sandro, a friendship which started with iron analyses in a chemistry lab. Sandro "seemed to be made of iron, and he was bound to iron by an ancient kinship: his father's fathers, he told me, had been tinkers (magnin) and blacksmiths in the Canavese valleys: they made nails on the charcoal forges, sheathed wagon wheels

---

[4] Review quotes from 1986 *Abacus* edition of 'The Periodic Table'

with red-hot hoops, pounded iron plates until deafened by the noise; and he himself when he saw the red vein of iron in the rock felt he was meeting a friend." Levi accompanied him on long, gruelling journeys to go skiing: "Sandro climbed the rocks more by instinct than technique...He seemed to feel he had wasted a day if he had not in some way gotten to the bottom of his reserve of energy..." In the end Sandro was killed "by a tommygun burst" fighting in a local resistance party against the fascists. A Mars character is etched, one who "lived completely in his deeds."

Are cobalt and nickel also of a Mars nature, as they have a close kinship with iron in many ways? This writer has ventured to claim such, but would Levi's tale support this? He first describes how close akin was nickel in rock to iron, and how it once resisted all his efforts to separate them: "we concluded that the nickel accompanying the bivalent iron took its place vicariously, followed it like a bivalent shadow, a minuscule brother: 0.2 percent of nickel, 8 percent of iron." After discussing the etymology of nickel in terms of the *cupfernickel* sprites which once lurked down the mines to deceive the miners, also the *kobolds*, another breed of mine-elementals, Levi typically produces a Martial piece of philosophising:

> But this is no longer the time for sprites, nickel, and kobolds. We are chemists, that is, hunters: ours are 'the two experiences of adult life' of which Pavese spoke, success and failure, to kill the white whale or wreck the ship; one should not surrender to incomprehensible matter...

Finally Levi reflects on how fortunate it was that the method of extracting nickel from iron ore which he attempted was not after all industrially feasible, because then "the nickel produced would have entirely ended up in Fascist Italy's and Hitler Germany's armour plate and artillery shells." Nowadays nickel and cobalt have vital military uses in high-temperature and high-stress alloys, as in jet engines.

A chapter on Silver describes its pure and sensitive lunar nature. A firm prepared X-ray plates with silver bromide as usual, and then found that despite all its ultra-clean precautions the plates were going wrong, and would not develop. The customers were complaining. Every conceivable factor was tested, and finally the problem was traced to a pollutant being poured into a river some distance away; its water, after being filtered and passed through an ion-exchange resin to purify it, was used to wash the overalls worn by the staff at the firm. Minute traces of pollutant were brushing off the overalls and marring the plates. Levi's silver chapter is a lesson on how careful one has to be with this most receptive and impressionable of elements.

Occasionally one finds 'occult' books giving rulerships to modern metals like zinc. Am I being a mere stick-in-the-mud traditionalist in failing to see a celestial correspondence for this element? Like aluminium or calcium, it is a mere earth-element, a constituent of common clay. Support for this view comes from the maestro's judgement about zinc:

> they make tubs out of it for laundry, it is not an element which says much to the imagination, it is gray and its salts are colourless, it is not toxic, nor does it produce striking chromatic reactions; in short, it is a boring metal.

In the midst of his stories Levi sketches the essential being of an element, for example 'the generous good nature of tin, Jove's metal'. An expansive, jovial character appears in the tin chapter: 'Emilio's father was a majestic, benign old man with a white mustache and a thunderous voice", and "Emilio's father looked so respectable and authoritative..." The lead chapter is set as a fictional narrative in a distant time, and the narrator who is prospecting for lead explains: 'if one goes beyond appearances, lead is actually the metal of death: because it brings on death, because its weight is a desire to fall, and to fall is a property of corpses, because it's very colour is dulled-dead, because it is the metal of the planet Tuisto, which is the slowest of the planets, that is, the planet of the dead. I also told him that, in my opinion, lead is a material different from all other materials, a metal which you feel is tired, perhaps tired of transforming itself and that does not want to transform itself anymore: the ashes of who knows how many other elements full of life...' (One thinks of all the radioactive elements whose decay-paths over millenia have terminated as lead isotopes.)

Mercury required a fictional tale, set on a strange island with a shifty mercurial character called Hendrik who tries to abduct the narrator's wife:

> [Hendrik] and Maggie took long walks together, and I heard them talk about the seven keys, Hermes Trismegistus, the union of contraries, and other rather obscure matters. Hendrik

built himself a sturdy hut without windows, put his trunk in it, and spent whole days there, sometimes with Maggie: you could see smoke rising out of the chimney. They would also go to the cave and return with coloured stones, which Hendrik called 'cinnabars.

One thing leads to another, and finally the narrator pins Hendrik to the wall at knife-point to get some sense out of him: 'Quite a tale this was, clear, straight talk, truly that of an alchemist, of which I didn't believe a word.' The lives of the characters transform in this tale as they each gain new marriage partners. One hopes that futurity will find industrial chemists coming to respect Hermetic lore a little more than is indicated by Levi's Mercury chapter!

5. Mutations of Mercury-1

by Tony Jackson

Mercury is the smallest planet in the solar system and the planet nearest the Sun. A scorching world inescapably close to the burning heat and shining brightness of the star that gives life to our solar system. So close to the Sun that it gives ear to the music of the Sun's awesome, nuclear, celestial process, the Music of the Spheres. Hence its role as messenger of the gods.

Throughout the history of astrology, Mercury's prime allegiance has always been to the Sun, as their physical proximity confirms. Myth also accords a solar influence, but by virtue of familial relationship. Hermes' kinship with the Sun was through brotherhood, the sun as Apollo. The Greeks endowed Mercury (Hermes) with an intellectual brilliance and a savage cunning unbecoming to moral virtue. As such, Mercury represents mankind more than any other planet in the astrological chart.

Mercury is so close to the Sun that it is virtually cannot be seen with the naked eye. In addition, Mercury's only sighting position from earth is such that only one face of Mercury ever shows to us.[1] Thus, as a metaphor, both astronomically and astrologically, if Mercury is taken at face value, it is *neither seen nor understood* - and, in essence, one must step outside the ordinary means of observing it in order to know of its full value.

Hard to grasp is a theme that runs through all of Mercury's manifestations. Not only astronomically and astrologically but also as the metal itself. The heavy, silver-white metal otherwise called quicksilver is commonly obtained from cinnabar, its most important metallic ore. Its chemical symbol is Hg (*Hydrargyrum*, 'liquid silver'). It is probably most commonly known for its use in thermometers as the column of mercury rises and falls with temperature changes and for its controversial use in dentistry. Absorbing other metals, it forms amalgams and is used for filling dental cavities - often with catastrophic results.

---

[1] See eg my http://astrozero.co.uk/astronomy/mercury.html

Mercury has a quite varied chemistry, whose applications range from photography to pesticides. Fluid at room temperature, it maintains its quality of being physically hard to grasp, as any child who has broken the thermometer has found out to his delight when playing with this funny, shiny, elusive, runny stuff that keeps trying to escape from under the fingers.

Mercurius was also the true object of the alchemical procedure. Quicksilver, because of its fluidity and volatility, was also defined as dry water or *Aqua sicca.* It is the very Spirit of Alchemy, the alchemical fire, the universal and scintillating fire in the light of nature which carries the heavenly spirit within it.[2]

In mythology Mercury was the Roman name given to Greek Hermes, son of Zeus and brother to the Sun-god Apollo. Hermes has his counterpart in Egypt as Thoth, son of Ra the Sun-god. Thoth was scribe to the Sun-god, but also advocate for the dead; this dual role was handed through to Hermes as messenger of the gods and in his

---

[2] C.G. Jung, *Alchemical Studies*, (Collected Works, vol. 11), 209

role as psycho-pomp. But otherwise they diverge as quite different characters: Hermes' link with Apollo is based on sibling rivalry, whereas Thoth's interaction with Ra shows a respect for authority. Hermes appears as a punk provocateur but Thoth behaves more as the humble servant. Hermes' inclinations for flight suggest rulership by Gemini, while Thoth's discriminating and honourable nature implies Virgo's influence.

In astrological terms Mercury is about connection, making links, communication, wits and intellect: quicksilver/quick mind, connecting things and ideas, revelation, the imagination, language and understanding. *Mercurius* plays with ideas and finds links between them. Above all, Mercury is about process, and that implies movement, conscious and unconscious, physical and non-physical - turning on the light, metaphorically speaking (Mercury has an industrial use in electric switches). Mercury rules connections, from roads and transport to telephone and radio, from information systems to the world-wide web and satellite communications (There is a telephone company called Mercury). Hermes rules roads and travel, protecting the crossroads, highways and byways. We can imagine a man travelling along the obscure roads of life, either guided or led astray by Hermes, and in need of an instinctual omen (Hermes) to mark his movement along the way.

Making connections is also about appropriate response, and needs efficiency and quickness of comprehension and mental agility. Someone who is well-connected to their Mercury has all these qualities, but when the ray (as Misha Norland describes it[3]) is broken and pathologised, Mercury turns into its opposite: mentally inefficient and slow to comprehend what is being asked of them, hurried, restless, impulsive and unable to focus, so he can lose his way in well-known streets and will answer questions incorrectly - the darker face of Hermes.

In the Homeric *Hymn to Hermes,* we learn for the first time of Hermes the trickster. While still only a babe in arms he stole his brother's cattle, and then "cut off from the herd the fifty loud-lowing kyne, and drove them straggling-wise across a sandy place, turning the hoof prints aside. Also, he bethought him of a crafty ruse and reverse the marks of their hooves, making the front behind and hind before,

---

[3] Misha Norland, *Spiritual Aspects of Homeopathy,* 2003.

while he himself walked the other way."[4]

In driving the cattle, Hermes simulates a backwards movement, skilfully done by reversing the sandals and the hoof prints of the cattle, a movement in one direction which appears as a movement in the opposite direction.[5] For me this represents and is a metaphor for man in the world. Man in nature. Man having to use his wiles to survive, to trick the gods, to buck his fate. To be able to make connections to his own advantage.

Revelation is on-going for human beings in a sea of constant change, especially in these contemporary times, and we constantly need to be able to respond to the moment in appropriate fashion: to be fluid, like Mercury, and find new ways of freeing ourselves from outdated dogmas.

The Homeric *Hymn* continues: "Then he wove sandals with wicker-work by the sand of the sea, wonderful things, unthought of, unimagined; for he mixed together tamarisk and myrtle twigs, fastening together an armful of their fresh, young wood, and tied them, leaves and all, securely under his feet as light sandals". The sandals of Mercury represent the natural activities of man - what man does by nature. His inventiveness and dexterity. His ability to survive in a dangerous and tricky environment, to think on his feet.

Another son of Hermes was bathing in a pool of crystal clear-water, naked, and was seen by the naiad Salmacis. She falls in love with his beautiful form and cried out 'He is mine', and casting off all her garments dives into the waters. She holds him fast against all his struggles, stealing reluctant kisses, fondling him and touching his unwilling breast, and wraps him in her embrace as a coiled serpent. She cries out to the gods that they should never become separated and in hearing her prayer the gods grant her wish and the two bodies become knitted together as a Hermaphrodite. "After bathing in that uncanny pool with the nymph Salmacis, the beautiful boy, resembling Eros, realises his new condition and feels himself weakened, his limbs enfeebled".[5] The hermaphrodite raises imagery of sexual ambiguity,

---

[4] http://ancienthistory.about.com/library/bl/bl_text_homerhymn_hermes.htm
[5] Rafael Lopez-Pedraza, *Hermes and His Children*, 1989, 2003, 59

so dominant a symptom in Mercurius.⁶

The section on Mercurius, in Jung's *Alchemical Studies,* has a mine of information.⁷ It opens with the tale of a spirit confined to a bottle,⁸ buried amongst the roots of an old oak tree. The son of a poor wood-cutter is passing by, and he hears a voice calling to him from the oak, 'Let me out, let me out!' So he digs down, and finds a well-sealed glass bottle. A spirit is trapped inside it, and the young man holds a dialogue with this spirit; this involves, not surprisingly, a tricky promise and a spot of deception. After all, the spirit is called by the name of the pagan God, *Mercurius.* Eventually, he releases it from captivity, and he in return is made wealthy as his iron sword becomes turned into silver.

Both the oak and the forest represent the unconscious. Someone had confined the spirit 'hermetically' in the bottle and hidden it, deep down amongst the roots of the tree, which themselves extend right down into the mineral kingdom. "Presumably a magician, that is, an alchemist, caught and imprisoned it."⁹ This becomes an allegory for our true selves buried deep in the unconscious core of our personality, which must be discovered, refined and liberated.

The Mercurial essence--the spirit confined to the bottle--is tricky, unruly, fiery and artful. The alchemists isolated the spirit from its surrounding medium by hermetically sealing it in the 'Vas Hermeticum,' a round bottle made of glass which represents the cosmos in which the earth was created. We could interpreted the story thus: the pagan god has become an 'evil' spirit, 'forced under the influence of Christianity to descend into the dark underworld and be morally disqualified. Hermes becomes the demon of the mysteries ... the demon of forest and storm; Mercurius becomes the soul of the metals, the metallic man, the dragon, the roaring fiery lion, the night raven, and the black eagle--the last four being synonyms for the devil'.¹⁰

Mercurius was confined to the bottle by the alchemists in order to

---

⁶ Salmakis was a naiad nymph who dwelt in a magic pool, whose 'strengthless waters soften and enervate the limbs they touch ...' She fell for the youth Hermaphroditos: Ovid, *Metamorphoses* Vol. IV
⁷ *The Collected Works of Carl Gustav Jung,* Vol. 13 Alchemical Studies, IV The Spirit Mercurius.
⁸ Grimm's Fairy Tale *The Spirit in the Bottle.*
⁹ Jung op.cit., 195
¹⁰ Jung opp.cit., 198

transform him. They believed that he was so bedevilled and shameless that all who wished to investigate him would fall into madness. Mercurius was the *prima materia* of the alchemical process, representing and stemming from the collective unconscious. Mercurius was understood as quicksilver (Hg) and as such was called *vulgaris* and *crudus*. *Mercurius philisophicus* was specifically distinguished from this but was conceived to sometimes be present in *Mercurius crudus*. It was the true object of the alchemical procedure. Quicksilver was also defined as "the water that doesn't make the hands wet".

Many treatises define Mercurius as fire, a fire highly vaporous in its nature; an invisible fire that works in secret, or the universal and scintillating fire of nature which carries the heavenly spirit within it. This related Mercurius to the *Lumen Naturae*, the source of mystical knowledge second only to the holy revelation of the scriptures. Once more we catch a glimpse of Hermes as the god of revelation.

Hermes' role as messenger of the Gods allowed him to visit Hades, the king of the underworld. Hermes is the only appointed messenger to Hades and as such provides guidance into the psychology of the underworld with its difficult, frightening and pathological components of our psyche. This sheds light on the role of Hermes as psychopomp,[11] the bridge between the worlds, the conductor of departed souls

Mercurius is many-sided, changeable and deceitful. He is said to be dual-natured, enjoying the company equally of the good and the wicked.[12] He is the twin, made of two natures, the giant twofold substance, the two dragons. There is common Mercurius and philosophical Mercurius. The *Aurelia Occulta* gives a graphic description of him:

> I am the poisonous dragon, who is everywhere & can be cheaply had. That upon which I rest, and that which rests upon me, will be found within me by those who pursue their investigations in accordance with the rules of the art (Alchemy). My water & fire dissolve & compound; from my body the green & red lion may be extracted. But if you do not

---

[11] Psychopomp: one who conducts souls of the dead to an afterworld, as Charon
[12] C.G. Jung, Collected Works Vol. 14, 1963, *Mysterium Conjunctionis*, VI, 491

have exact knowledge of me, my fire will destroy your five senses. A most pernicious poison issues from my nostrils, which has wrought destruction upon many.[13]

From accounts of poisoning eg from mercury amalgam in teeth-fillings, we may gain a sense of "a most pernicious poison ... which has wrought destruction upon many." The *Aurelius Occulta* continues:

> By the philosophers I am named Mercurius; my spouse is the philosophic gold; I am the old dragon, found everywhere on the face of the earth, father and mother, youthful and ancient, weak and yet very strong, life and death, visible and invisible; I descend into the earth and ascend to the heavens, I am highest and lowest ... I am the carbuncle of the Sun, a most noble, purified earth, by which you may turn copper, iron, tin and lead into purest gold.

Thus we see the dual nature of Mercurius and in so doing can see how representative he is of man on earth who also is dual-natured, capable of the highest revelation and most honorable deed as well as the foulest, most hellish acts. When connected to the Mercurial ray, man is 'quickened' and when not, he is dull. Mercurial man has his roots in the deepest depths and his branches reach up to the heights.[14]

Mutations of Mercury – II

Mercury appears sometimes in the form of a fluid metal, sometimes in the form of a hard brittle metal, sometimes in the form of a corrosive pellucid salt call'd Sublimate,

---

[13] Georg Beatus, *Aurelia Occultae Philosophorum* ('Basil Valentine'): *Theatrum Chemicum* 1613 Frankfurt, Vol IV, 569; 1659, 401: "Draco ego sum venenatus." (under the heading 'Materia Prima'). www.levity.com/alchemy/vaughan1.html Jung, op.cit., 218.

[14] Pictures by kind permission of the Wellcome Library, London. Tony Jackson was a jazz musician and his partner Sue Rose an astrologer, they must have had some interesting conversations to produce this. He gave me a copy of it before he died, and I tidied it up for Adam McLean's *Alchemy* website. This is a shortened version. My partner, who says she is psychic, told me that while she was kindly typing out this text, she had Tony Jackson jumping up and down on her bed: Hermetic philosophers may appreciate this image.

> *sometimes in the form of a tasteless, pellucid white Earth, call'd Mercurius dulcis, or in that of a red opake volatile Earth, call'd Cinnabar; or in that of a red or white Precipitate, or in that of a fluid Salt; and in distillation it turns into a Vapour, and being agitated in vacuo, it shines like Fire. And after all these Changes it returns again into its first form of Mercury.'*
>
> Sir Isaac Newton[15]

THAT PARADOXICAL ELEMENT, quicksilver, expert in transformations, affords many a classroom demonstration to delight the eye. In times to come a school chemistry teacher may well use the affinity which metals have with their parent planets to bring wonder and fascination into inorganic chemistry, a subject which can otherwise consist merely in the memorising of diverse facts of little interest to children. Hermes' metal mercury has interlinking properties, in that it associates with very diverse elements and amalgamates the metals together. Here are some classroom experiments which can reveal this (Caution - mercury salts are toxic):

* Adding a solution of potassium iodide to mercuric chloride solution turns it red, but if more is added then it mysteriously goes clear again. This is called double salt formation, because first mercuric iodide is formed and then potassium mercuric iodide develops which is colourless.

* The explosive mercuric fulminate may be prepared by heating a mercury salt with nitric acid and alcohol. This compound, made of dissimilar and conflicting substances, will explode when dry upon touching.

* The 'pharaoh's serpent' trick uses mercury thiocyanate. This compound is moulded into a paste with a cone shape and, once dry, when the top is lit, a huge serpent coils slowly out of it. Complex polymerisation is here taking place, between carbon and mercury.

* What chemists call sublimation can well be shown with mercury. This is a process whereby matter transforms directly from the solid into the vapour state without passing through the liquid condition. The red oxide of mercury is heated in a retort together with some common salt, until long, translucent needles of mercury chloride are seen to grow down from the top of the retort.

---

[15] Isaac Newton, *'Optics'* 1717, Query 30

* In vapour form, mercury glows ultra-violet when electricity is passed through it, and is still the commonest substance used in fluorescent tubing. The tube is painted with a substance which glows with white light when ultra-violet light strikes it.

* A reversible reaction can well be shown with mercury in preparing oxygen, using the same method a Joseph Priestley once did. On moderate heating, mercury will form its red oxide. If this oxide is then heated in a retort, it will decompose to give oxygen gas which can be collected.

A lesson about mercury could start by observing the shimmering motion of its globules in a bowl, as they form and reform. One could endeavour to lift up the stone jar in which mercury is kept: it always seems odd that a liquid should have such a very great weight. It lives in such a jar rather than glass or metal like other laboratory chemicals because it would swiftly escape from such, by breaking the glass or by dissolving the metal. Should mercury be spilt on the floor, the teacher will probably sprinkle suphur upon it to stop its toxic vapour from evolving, whereupon the yellow powder will be seen to transform into the reddish hue of cinnabar.

The derivation of the word, 'mercury' comes from the latin, *merc* or *merx,* meaning trade. Why should the name for a liquid metal come from a word meaning trade? The metal mercury is the one element one normally sees in three states of matter - as the fluorescent lamp overhead in the classroom, as the liquid in the thermometer and as calamine the skin lotion: as Hermes was the one deity who could come and go through the three worlds?

A Catalyst

Mercury is widely used in industry as a catalyst - either as itself, a metal, or in complicated organic compounds. When Hermes coupled with Aphrodite the result was a 'hermaphrodite', and likewise mercury gets involved in unlikely combinations. A catalyst helps to bring about a reaction while remaining itself unchanged at the end of the reaction.

The petrochemical industry uses mercury metal as a catalyst in its 'cracking' towers, to break down the heavy oils into a size that will burn in motor cars. As seems appropriate for a deity of speed and communication, Hermes' metal catalyses the formation of oils at just the octane (molecule length) required for the internal combustion engine.

Complicated 'organometallic' mercury compounds catalyse the synthesis of a range of pharmaceutical and other organic, man-made products. These catalysts, developed in the 1970s, are tricky to prepare and have to be formed at low temperatures, otherwise they tend to explode.

The most characteristic chemical trait of mercury is association. It links itself up in the most unexpected ways. A chemical textbook says of this metal, 'The tendency to form complex compounds is very marked in the case of mercury.'[16] It combines with nitrogen and carbon compounds which metals normally won't touch, for example mercurammine compounds, as well as forming the usual metal salts. In amalgamating other metals together it is also performing this interlinking function.

Some Mercurial Inventions

The metal mercury, which likes to be in constant motion, is used in the barometer to register changes in air pressure. Its height of rising reflects the pressure of the atmosphere. This was first demonstrated on the morning of September 19, 1648, in France. The mathematician Pascal had written to his cousin Francois Périer suggesting that he looked at the matter. So at 8a.m. Périer and some colleagues left one barometer at the foot of a high mountain in the Auvergne region of France, and climbed up to the top with another barometer. Appropriately enough, at the top was located a ruined temple to Mercury. It was seen that the level of the mercury at the top of the montain was definitely lower than at the bottom of the mountain. To quote James Burke, 'Everyone was elated. The barometer had been invented.'[17] It probably took about three hours to climb the mountain, which puts them at the summit at 11 a.m.. At his time, Mercury had just culminated i.e., reached its highest point, and was in close square to the three outer planets Uranus, Neptune and Pluto, which seems quite appropriate.

---

[16] J.R.Partington, *'A Textbook of Inorganic Chemistry'* 1960 p.795.
[17] J. Burke, *'The Day the Universe Changed'*,1985, p.75.

Giorgio Piccardi, formerly chemistry professor at the University of Florence, noticed that certain chemical reactions could vary in rate because of radiation from space. He was able to show that a slow colloidal precipitation reaction (using bismuth oxychloride) would vary in rate with the activity of the Sun, with solar flares and with the whole 11-year cycle; and it also varied with the phase of the Moon. Beginning in 1951, many thousands of such measurements were made over the years to ascertain this, and the Piccardi Institute in Florence now continues this research.[18] Piccardi concluded that the structure of water could be altered by the radiation from outer space, and this was the reason why his chemical reaction rates were different from one day to the next.

Piccardi's interest in this matter began in a curious fashion. He had in Florence a method of descaling boilers. It involved what he called 'activated water', whose recipe he had devised. Scoff as his colleagues might, Piccardi kept being called in to descale boilers, using his special water. His recipe for the 'activated water' was unusual:

> A vial of glass containing a drop of mercury and low pressure neon is slowly stirred inside the water. As the container is moved, the mercury rubs against the glass; the double electric layer between the mercury and the glass breaks and produces a red luminescent discharge through the neon. The water touching the vial ends up activated.[19]

Here, one feels, was a man who appreciated the mystery of mercury. Indeed, there is something alchemical in the approach. He claimed that the efficacy of his mercurial brew varied with the influx of solar radiation, and thereby was led onto his better-known colloid precipitation experiments.

In the birthchart of Giorgio Piccardi (born Florence, Italy, at 23.30 local time 13/10/1895) Mercury was conjunct Saturn, which is certainly appropriate for scientific work, and was also in square to Jupiter. From the mundane affairs of boiler technology, the mind of this bold Florentine reached out into the depths of space, and forged a new theory as to how earth and sky were interlinked.

---

[18] G.Piccardi, 'The Chemical Basis of Medical Climatology,' 1962. For an account of Piccardi's work, see Gauquelin, M. 'The Cosmic Clocks', 1969 or 'The Cycles of Heaven', Playfair & Hill, 1979.
[19] Quoted in Gauquelin, Op. Cit.,p.175.

The first electronic computers used mercury for their short-term memory, to hold the electronic information. Built in the 1940s, they had 'mercury delay tubes' to store electronic pulses. Speedy mercury was clearly just the element for this task. To quote from a book about the first computers, as distinct from mere calculating engines, built in the mid-1940s: 'This appropriate element, associated with the classical deity of speed and communication, was to haunt the developments of the next few years.[20]

Mercury in Alchemy

The metal mercury had a very special role in alchemy; however as the language of alchemy was somewhat obscure it may be that we will never understand this. The earliest alchemic documents, from the third century AD, are shortly after the first account of the preparation of quicksilver in the second century AD, so perhaps the discovery of mercury produced it. The Indian word for alchemy is 'Rassayana' which means 'the way of mercury.' A recent work 'The Chemical Theatre' by Maurice Nicholl has well explored the use of alchemic themes in Elizabethan drama and Shakespeare in particular. Let us quote from it:

> 'Like all the metals most familiar to the early chemist, Mercury bore the name of a god. The metal was invested with properties associated with the god. Mercury's liquidity as chemical corresponded with the nimbleness of the god, just as iron was strong like Mars and lead old (decadent) like Saturn. The more esoteric applications of Mercury - as a volatile spirit, ascending and descending anima - related to Mercury as winged messenger between gods and men. The dissolving aspect of alchemical Mercury links with the mischevous character of the god Mercury, patron of rogues, vagabonds and pickpockets... the alchemists' exasperation must be seen in the context of their reverence for Mercury as the secret and agent of transformation... Someone who is mercurial is (according to the Oxford dictionary) 'sprightly, ready-witted and volatile', and

---

[20] Andrew Hodges, *The Enigma of Intelligence*, 1983, p.315.

(according to Roget) 'changeable, mobile, quick, excitable, elusive.'[21]

In many guises the mutable and tricky figure of Mercury appears in alchemic texts:

> *Alchemist*: Do but tell me if thou art the true Mercury, or if there be another.
>
> *Mercury:* I am Mercury, but there is another.[22]

This was in a work published in 1605. The previous year a play 'Eastward Hoe' appeared by Ben Johnson dealing with alchemic themes, and with characters having suggestive names like Golding, touchstone and Quicksilver. Quicksilver attempts to corrupt Golding him and corrode his resolution, as mercury amalgamates with gold and thereby rob it of its strengh. The drift of the play follows the fixing, evaporation etc of quicksilver, who is described in terms like, 'Whither art thou running?', 'mad Quicksilver', 'my nimble-spirited quicksilver' and 'my runagate quicksilver'. When finally he is led to the gallows, Touchstone gloats, 'There is my quicksilver fixed.' In 1610 Ben Johnson's satire 'The Alchemist' appeared, where Mercury was characterised (by Face) as 'A very fugitive, he will be gone.'

Ben Johnson displayed a total scepticism towards alchemy, to the extent that when his library was destroyed by fire in 1623 a friend wrote that undoubtedly this was 'a mere act of retaliation on the part of Vulcan.' Of the two contemporaries Shakespeare and Ben Johnson, the latter treated alchemy with bitter scorn and derision, while the former always introduced it with deep reverence, for example:

> *Full many a glorious morning have I seen...*
> *Gilding pale streams with heavenly alchemy.*

Few persons today, wondering whether anyone prior to Lord Rutherford managed to transform one element into another, are likely to be convinced by the arcane language of alchemy. Before the sober gaze of the modern chemist, the strange dreams of the alchemists appear as mere vapours devoid of real form. Yet it was none other

---

[21] C.Nicholl, *'The Chemical Theatre'*,1980, p.179-80. Further quotes in this section come from this work.
[22] Michael Sendivogius, *Novum Lumen Chemicum,* Prague 1604, p67. (Op. Cit, p.183)

than Robert Boyle whose influence led to the repeal of the law prohibiting transmutation (Chapter 8). His biographers give curiously little emphasis to this historic deed, which arguably signified the demise of the alchemic tradition as much as anything else.

We turn now to a view of mercury by a modern industrial chemist and poet - an unusual combination - the Italian Primo Levi. His biographic work 'The Periodic Table' finds him grappling with the elements one by one, each one recalling some incident in his life*. In these chapters, the planetary archetypes rumble in a rather impressive subterranean manner. His Mercury chapter tells of life on an island. Of a mercury-figure 'Hendrik' we learn: 'He had a crooked, fleeting light in his eye like that of Mercury.' This is a look askance, indeed. Finally the author gets an explanation from 'Hendrik':

> Mercury, for their work, would be indispensable, because it is a fixed volatile spirit, that is, the female principle, and combined with sulphur, which is hot male earth, permits you to obtain the philosophic egg...in it are united and commingled male and female. Quite a tale was this, clear straight talk, truly that of an alchemist, of which I didn't believe a word.

Phantom Red Mercury

The film 'Red Mercury' described how a terror-cell was discovered just in time, stopping its fiendish plot to use a deadly new bomb that used 'red mercury.' It was released just before the London bombings of July 7th, 2005, so the timing was eerie.

A year later, in London's venerable Old Bailey courtroom, three men were cleared of trying to procure the raw ingredients for such a "dirty bomb" - which, the prosecution claimed could have devastated a British city if it fell into the hands of terrorists. So what exactly is this stuff?

It may not exist. The prosecutor for the crown warned the jury not to get "hung up" on whether red mercury actually existed at all:

> The Crown's position is that whether red mercury does or does not exist is irrelevant.

The Mad Hatter would definitely have a comment here. Red mercury occupies a liminal space in the outer fringes of modern conspiracy theory. It emerged during this trial that red mercury was something of an urban myth, a substance which was either radioactive or toxic or neither, depending on who you spoke to.

Was it maybe "an alpha crystalline form of mercury iodide, which changes to a yellow colour at very high temperatures"? It was said to change hands at prices of up to $300,000 per kilogram on the black market. Was it maybe connected to nuclear fusion technology, for use in small, portable bombs? In September 2004 the *News of the World* claimed to have foiled a fiendish terrorist plot to buy red mercury as material for a dirty bomb, reporting an exchange price of $300,000 a kilo. But according to a 1994 investigation by the Russian prosecutor-general's office, the "red mercury" being sold by Russian conmen throughout Europe and the Middle East (after the collapse of the Soviet Union) was merely mercury dyed red with nail varnish

.The conspiracy site 'Rense' quoted the inventor of the neutron bomb, Sam Cohen, deploring "the relative ease by which neutron bombs can be created with a substance called red mercury." But thank goodness, The International Atomic Energy Authority in Vienna had a clear view of the matter: "Red mercury doesn't exist," a spokesman said. So, will the story now lie down and stay dead?

No chance! Mercurius the trickster will be hatching some fiendish new plot even now...

Anti-Grav?

You'll be surprised how many people believe that anti-grav levitation technology has been developed by the Secret Government by back-engineering crashed UFOs etc - however they are not about to tell us. Such designs will most often tend to use speedy mercury, just as did the Vimanas or flying craft of ancient India. Does it maybe use high-speed rotation of some mercury-type plasma, even supercooled, exotically glowing violet with electrical beams?

I have a dream, of a school chemistry lesson, which would allude to the marvellous Indian stories of the *Vimana*, by way of promoting a multicultural approach to the history of mercury – but would *not* comment on whether they 'really' existed or worked. *Au contraire* pupils would be invited to make such judgements themselves. It would be explained that chemistry has a mythic side to it, and this is a part of that. The point of the story here, is to show the subtle and marvellous powers attributed to mercury, by way of levity, which is the opposite of gravity. Every Indian knows the story in the classic *Ramayana,* where Rama rescued his partner Sita from Ceylon, carrying her off in his flying car, to Ayodya in central India. The sole responsibility of the teacher here, is to ensure that decent classical references are here used, not just web-gossip.

An 11th-century Indian text the *Samarangana Sutradhara* is descrived by Wikipedia as 'an encyclopedic work on classical Indian architecture'. Amongst its eighty chapters on all sorts of topics, its chapter 31 concerns 'Preparation of Mechanical Contrivances'. This starts off by explaining (according to a new 2007 translation[23]):

> Now we shall dilate upon the chapter on Mechanical Contrivances come into order as per dilation which constitutes the sole source of Dharma (spiritual good) Artha (material gains), Kama (lust) and Moksa (final emancipation).

Translation from ancient Sanskrit into English is never going to be easy. The four elements of matter are described, plus ether the fifth:

> The constituent element of that may be of four kinds viz. The Earth, the Water, the Fire and Air owing to the medium of action of these ones even the 'Vijayat' (ether) is also put into usage. By which ones, Suta or mercury has been declared as a distinct (element) they do not know it correctly. By nature Mercury is Parthiva i.e.to be brought under Earth...'

Then, rather too briefly, the middle of this chapter gives an unlikely account of how to construct a *vimana*, a flying craft. This has a 'lost in translation' air:

> Having created a huge bird of lighter timber having frame work well knit and stout (one) may introduce *rasayantra* (i.e. mercury contrivance) having the base for fire. Mounted over that the man by the breeze released by the flapping locomotion of the wind or flank twain of that one, having inner self cosy by the power of mercury spelling marvel, flies along afar.

Here is a possibly more helpful translation of the Sanskrit:

> Strong and durable must the body be made, like a great flying bird of light material. Inside it one must place the mercury engine with its iron heating apparatus beneath. By means of the power latent in the mercury which sets the driving whirlwind in motion, a man sitting inside may travel a great distance in the sky in a most marvellous manner.

Here the 'driving whirlwind' is said to allude to some sort of mercury turbine engine. To continue:

---

[23] Vol. 1 of *Samarangana Sutradhara* translated by S. Sharma, Delhi, 2007, pp.363-282.

> This very way similar in extent to a gods' abode or temple; flies along a heavy aerial car or vehicle (made) or timber or wood. One may provide in an orderly way on the four corners of that one, stout ewers mercury-laden.
> By the power of rasaraja (quicksilver or *parada* i.e. Mercury) heated and exploding quickly ((Jhagiti) it becomes an object of decoration in the sky owing to the quality born of the heated pots by slow ignition burnt within the steel or Iron potsherds.

Again, the following translation may be more helpful:

> Four strong mercury containers must be built into the interior structure. When these have been heated by controlled fire from iron containers, the *Vimana* develops thunder-power through the mercury. And at once it becomes like a pearl in the sky.[24]

NB, Rasāyana, is a Sanskrit word that literally means Path (*āyana*) of essence (*rasa*). It is a term that in early ayurvedic medicine means the science of lengthening lifespan, and in later (post 8th-century) works sometimes refers to Indian alchemy. The name of the science of Indian alchemy is more generally "The Science of Mercury", or *Rasaśāstra,* in Sanskrit (Wiki).

The other great Indian epic the *Mahabarata* has floating cities in the sky with craft a-buzzing around - until a dreadful war put an end to it all. Both of the great Indian epics feature flying 'cars.' The *Ramayana* was composed in the 12th century, a century after the 'Mechanical Contrivances' text we've just looked at, concerning mercury-powered levitation!

Up a Filter paper

No-one is interested in the work of Agnes Fyfe these days, that's for sure. But, allow me here to dredge it up briefly from oblivion. Some decades ago, she worked at the Weleda Clinic at Arlesheim in Switzerland, a naturopathic cancer-cure centre: their remedy used

---

[24] 11. *Atlantis and the Power System of the Gods, Mercury Vortex Generators and the Power system of the ancients,* by D.H. Childress and B. Clendenon, p.91.

mistletoe, and her time-experiments studied how its vitality would vary with cosmic cycles of time.

She conducted experiments of an *elementary* nature using extracts of plant sap. She would take a small amount of plant extract, suitably diluted and rise it up a filterpaper as a chromatogram, then let it dry, then twice the volume of metal salt solution (usually silver or gold) at 1% concentration would be risen up through it, alowing characteristic images to develop on the paper. She did experiments of this kind at Arlesheim over some three decades. She was employed by a Swiss cancer clinic which uses a remedy prepared from mistletoe in the treatment of cancer.

She believed that the *time* at which the mistletoe is picked was of significance for its healing virtue. Over many years Fyfe has claimed that eclipses, in particular, wipe out the normal energy of plants and that chromatograms made with the sap and metal salt solutions could show this:

> Total lunar eclipses for example, which occur comparatively frequently, always bring about the same kind of change in the general configuration of the plant pictures.[25]

The 'plant pictures' ie chromatograms here were from plants picked during the eclipse: Fyfe believed that it was the virtue of the plant at the time of *picking* which the chromatogram depicted, and in this respect the experiments differed from those performed solely with metal salt solutions, where the time that is supposed to matter is when the solutions are mixed to rise up the filterpaper. The sap remembers as it were the time it was picked: the surging ebb and flow of life-forces in the plant, mirroring the changing sky, was supposed to halt in the plant once severed from Mother Earth. This rather romantic picture is what Fyfe claimed to have found depicted in her chromatograms.

Fyfe's last work was about Mercury. A special Mercury-effect showed up during Sun-Mercury conjunctions she claimed, and its nature could be simulated by adding a small amount of mercury salt to the gold reagent solution. For example over a total solar eclipse, using sap from iris as well as mistletoe plants, she found that:

---

[25] 'The Signature of the Planet Mercury in Plants', Part III, Agnes Fyfe, *British Homeopathic Journal*, 1974 Vol.LXIV, p.111 (trans. from German).

> Both plants showed clearly, with neutral pictures, the time of the Mercury/Moon conjunction which occurred just before the middle of the eclipse, but only mistletoe reacted definitely to the conjunction of mercury with the Sun.[26]

This seems to be suggesting that mistletoe is of a solar virtue (as the Druids of old believed) whereby it will respond especially to solar events.

She described how different metal solutions could be risen up through plant sap in a filterpaper:

> The gold reagent gives rose colours, the silver nitrate shades of brown, the copper combinations produce slightly differing greens: the mercury bichloride gives no colour at all, as if it were water, but the slight colour imparted by the sap shows exactly how high the liquids rise.[27]

Thus the different metals express themselves - while elusive mercury gives nothing away. (Iron solution gives a reddish-brown hue.) One may be reminded of the statistical experiments of the Gauquelins: they discovered planetary links to the birthcharts of eminent professional groups, for example that the Moon helped imaginative writers and Venus promoted artists or musicians, while doctors had the aid of stern Saturn, but with Mercury they found nothing - its liquid essence would not be caught by their webs of statistics.[28]

Such experiments are pleasing for children, especially using gold where fine 'auroras' of colour effects are obtained, as befits its solar essence. Next we come to a claim which is rather odd: Fyfe alleged that the height to which the four metallic solutions rose after passing through the plant sap, ended on when they were done. Sometimes the copper and mercury rose higher than the silver and gold while at other times it was the other way round. These vertical chromatogram experiments were done in Fyfe's laboratory where room temperature and humidity were kept constant (20°C and 70% relative humidity). Fyfe measured the height of each filterpaper to which each reagent had risen and plotted these on a day-to-day basis. She claimed that the

---

[26] Ibid, Part I (Jan. 1974), p.58.
[27] Ibid, p58.
[28] For my review of Gauquelin's work, see 'How Ertel Rescued the Gauquelin Effect' in *Correlation* (online).

mercury rose higher than the gold solution over each inferior conjunction of Mercury with the Sun, ie when Mercury moved in between Earth and Sun, but that normally it was the other way round. Superior conjunctions, ie when Mercury was on the far side of the Sun from Earth, did not in her view have so much effect. Such an experiment might not be too difficult to repeat, as it may not need very uniform temperature conditions, involving merely the relative height of rising of one solution compared to another.

I was once advised by Fyfe to use plant sap experiments in investigating celestial events, on the grounds that they responded better than mere metal salt solutions which I had been using, however rightly or wrongly the advice was not taken up. Sap viscosity after extraction from a plant is by no means easily standardised, and a temperature and humidity-controlled lab such as Fyfe enjoyed is required to help ensure that the sap rose under uniform conditions.

Fyfe's work, her study of celestial influence upon herbal virtue, is usually criticised for being not sufficiently quantitative and therefore hard to interpret. This is indeed its main problem: one ends up with thousands of chromatograms, and presumably the person who conducted the experiment has been convinced, but has the information been expressed in a publicly accessible form? Reproductions of selected chromatogram pictures in books or journals by themselves tend not to be very convincing, and Fyfe's life-work has faded into oblivion for this reason. We continue in the next chapter with her experiments using copper.

'A Son of Many Shifts'

Hermes became associated with the planet Mercury in the fourth and third centuries B.C., as the Babylonian star-lore infiltrated into the Hellenic world. Thus Plato refers to the planet Mercury as 'the one called sacred to Hermes.'[29] But originally the Greek deities were not astral deities as such, in contrast with those of Chaldea. Before this, the traditional Greek name for the planet Mercury just meant 'the scintillating one'.

---

[29] 'Plato' Volume IX, Loeb classical Library, 1975, p.79.

In Homer's tale, we find Apollo addressing Hermes as 'You comrade of dark night', after his shocking behaviour of stealing Apollo's cows, adding further, that he is

> a son of many shifts, blandly cunning, a robber, a cattle driver, bringer of dreams, a watcher by night... [30]

All ends well however, when Hermes presents Apollo with a lyre, saying:

> From now on bring it to the rich feast and lovely dance and glorious revel, a joy by night and day. Whoso with wit and wisdom enquires of it cunningly, him it teaches through sound all manner of things that delight the mind.

---

[30] 'The Homeric Hymns & Homerica', Translated by H.G.E. White, Loeb Classical Library, 1977, p.399.

6. Aphrodite, Copper and Venus

"My song shall concern Aphrodite,
In her crown of gold, the beauty, protecting each
High town of Cyprus which is in the sea:
There the moist breath of the West Wind conveyed her
Over the sounding swell of the sea
In the softness of foam, there the gold-circleted
Hours welcomed her gladly, dressed her in clothes
Of the gods, and placed on her undying
Head that crown of the finely-worked gold,
And hung her pierced ears with earrings
Of splendour made of the gold-coloured
Copper out of the mountains..."
Homeric Fragment[1]

'Copper, see Venus' says the index of a work on the history of alchemy - and quite right too![2] Nowadays the question should be formulated as, how do the modes of function of this metal reflect traditional qualities associated with the planet Venus? It is evident that such a question does not refer to the attributes of a rock-strewn planet surrounded with boiling sulphuric acid vapour - but rather to the traditional image associated with it, of a love-goddess. Thus, the word 'venerate' derives from Venus. With a perhaps comparable (though equally mysterious) etymology, the word 'desire' derives from 'de-sidera', from the stars. Such an imaginative image associated with the planet Venus represents a certain triumph of the imagination over physical fact, in a way that

---

[1] Homeric fragment, quoted in *The Goddess of Love* by Geoffrey Grigson, 1978 p45.
[2] Charles Nicoll, *The Chemical Theatre*, London 1980 p.287

can leave one uneasy. A study of the metal associated with the planet may help us to fathom this difficult issue.

Copper is a sensuous and attractive metal - as Mellie Uyldert writes, 'Bring back copper and treat it with respect. Let us see again the kettles, chandeliers, pans, lamps, tobacco tins, extinguishers and taps so that we can polish them, touch them and look at them!'[3]

This advice is well followed in British public houses, which thereby achieve a convivial atmosphere. Our image of beauty tends to be associated with a copper tan, or a well-bronzed figure. (The skin pigment responsible, melanin, requires copper to form the tan) The radiant hues of a peacock's plumage are woven from copper complexes, and many a bird's wing iridesces with the green-blue hues of copper compounds.[4] At the temple of Aphrodite at Aphrodisias in what is now southern Turkey, a figure of the goddess cast in copper was found.

Figure: an ancient Greek head of Aphrodite, from Aphrodisias

---

[3] Mellie Uyldert, *Metal Magic*, London 1980, p70 (translated from the Dutch)
[4] W. Pelikan, *The Secrets of Metals*, 1973 New York, pp110-111 (translated from the German).

Many of the pictures here shown need to be viewed in colour, and you can get the kindle edition for this. Indeed I have here omitted many of the images, as having little point in black and white.

The well-known picture of *The Birth of Venus* by Sandro Botticelli reminds one of the green-blue marine hues of copper stones and compounds. The background in the picture - painted using copper pigments[5] - shows the characteristic green-blue hue of copper chloride or sulphate, the same colours as appear looking through fine copper leaf, and when copper is vapourised. When copper filings are sprinkled over a flame, similar colours are seen. The skin of Venus-Aphrodite in the Botticelli painting is given a delicate bronze hue and the hair is copper coloured. The scallop shell in the picture, traditionally associated with Venus, comes from a sea-creature which respires using copper instead of iron, as does the octopus, and this gives a slower metabolism.

Tourist books about Cyprus describe the two main subjects of note regarding the island's history - the copper mines which put it on the map, and the birth of Aphrodite on its shores - without ever noticing a connection between them. The story would be a good deal more interesting if it were appreciated that the island of Cyprus, in its geology and its mythology, combines the microcosmic working of the planet Venus in its copper veins with the macrocosmic image of Venus in its legends. That would stimulate tourism to 'this rich island of flowers, corn and copper.'[15] Certain copper salts used to be called 'cuprous', as in cuprous chloride, the term having the same root as Cyprus (*Kyprus*).

The author was once attempting a copper crystal experiment over an occultation of Venus with the Moon. The date was July 15, 1974 at the New School, King's Langley. An interested schoolteacher had arranged for his chemistry class all to try the experiment by dropping copper sulphate crystals into silica water. Would they grow differentially over the occultation? This is the well-known 'Chemical garden' experiment.

While the experiment may not have been a great success, geopolitical events of that day told a story. Turkish troops invaded Cyprus to abort a plot by the colonels' junta in Athens to overthrow Archbishop Makarious and bring about Enosis - union with Greece.

---

[5] Grigson, op.cit., p.28.

The Partition of Cyprus dates from that fateful day in 1974. Archbishop Makarious then *disappeared* from public view as his palace was shelled and was presumed dead - over the very hours when Venus was occulted by (ie, went behind) the Moon - and then he *reappeared* in the evening at the port of Paphos, to quit his country. Yes, that's when it happened!

What follows could be read merely as an account of how a specific metal functions in the human female metabolism; one hopes however that readers will use their imagination to perceive in this account the Venus-principle working in copper. It will indeed strike a *venereal* note, and we may here consider that *venerate* and *venereal* both stem from the same Latin root - Venus.

Copper/Venus in human metabolism

Copper in the female metabolism exhibits some remarkable Venus-characteristics. Elsewhere the author has described how other metals function in human physiology in accord with traditionally-assigned planetary characters. Iron, with which copper is so closely related in many of its functions, colours blood red with the iron-containing molecule haemoglobin, as it also colours the planet Mars red with the oxide of iron. Here the symbolism is of a bold and unmistakable nature, whereby the Mars-energy of iron is linked to warfare and strength of will - lack of iron in the blood causes anaemia, a condition associated with weakness of the will. By comparison, 'Saturnism' is the traditional name for lead-poisoning, and this metal accumulates in the skeleton, part of the body traditionally associated with Saturn. In the case of copper, the metal occurs in much lower concentrations than iron in the human body, in a total quantity of a mere one-tenth of a gramme, and therefore we find its planetary affinity taking more subtle modes of expression.

Biochemically the level of copper in the blood is critical, being around one part per million by weight, and normally it remains fairly steady around this value. On average, women have about 20% higher copper serum than men and for iron it is the other way round,[6] with men having a one-third higher iron level than women in their blood.

---

[6] Schenker et.Al.,'Serum Copper Levels in Normal and Pathologic Pregnancies,Am.J.Obs.&Gyn.,1969,105,pp933-937.

The significance of this fact is entirely ignored by modern medecine. Iron and copper levels are sex-linked in exactly the way expected from the gender symbolism of their planets.

While measuring elements in the human placenta at birth, scientists were surprised to notice that just one of them was strongly sex-linked.[7] it was appearing at much higher levels in the placenta around newborn girls than for boys, and this element of course was copper. Somehow, this metal has a uniquely feminine function in human physiology, and it is time biologists asked themselves, why this should be. A female foetus is surrounded by a sphere of copper, has a higher concentration of the Venus-metal around it, helping it to become a girl.

Copper in women's blood serum has a monthly cycle in tune with their menstrual period, peaking a week or so before the period arrives.[8] This is because their serum copper exists chiefly as the protein, 'ceruloplasmin', whose metabolism is closely linked to the female sex-hormone oestrogen.

Normally physiologists refer to only one function of copper: ceruloplasmin in blood serum assists in the production of the iron-molecule haemoglobin, within blood corpuscles. Mars and Venus were closely linked in mythology, and in human blood their two metals work in a partnership. And yet, it would be a mistake to omit the subtle ways in which copper acts in its own right, merely because they are not yet well understood. For example, some doctors now treat pre-menstrual tension by reducing body copper levels. A mild chelating agent is administered for this purpose (This is a medicine which flushes out heavy metals from the bloodstream).[9]

The Brain-Bio Centre in Princeton, USA, founded by Dr Carl Pfeiffer, has treated several thousand such cases in this way, with positive results.

---

[7] 'Copper was the only element for which highly significant sex differences was found (higher in female neonates) 'Ward, Watson and Bryce-smith, Placental element levels in relation to fetal development' *Int Jnl for Biosocial Res.* 9, 1987, p.68. .

[8] W. von Studnitz and D.Berezin, 'Studies on Serum Copper during pregnancy,during the Menstrual Cycle and after the administration of Oestrogens',Acta Endocrinologia,1958,27,p245-252. S.Dokumov,'Cuivre Serique et Cycle Menstruel',Rev.Fr. Gynecol. Obstetric.,1968,63,p37-42.

[9] The treatment of PMT by copper serum reduction at the Princeton Brain-Bio Centre was confirmed by Dr Pfeiffer in a letter. Vitamin C and zinc treatment are prescribed, and possibly also a chelating agent.

Unfortunately, copper can nowadays reach unduly high levels in the serum, especially in women taking the contraceptive Pill, and then this excess copper becomes associated with symptoms of stress. Water pipes are usually made of copper, its pliable nature making it suitable for plumbing, so drinking water can contain relatively high levels in solution. The Pill works by emulating conditions of pregnancy where oestrogen is high, and this has a drastic effect upon serum copper levels.

During pregnancy, copper serum in the mother climbs up to double its normal level, reaching 1.9 parts per million. Conversely, iron in foetal blood also increases as the time of birth approaches,[10] so a copper-iron polarity develops between mother and child. Insomnia, depression and changeable moods towards the end of pregnancy have been related to the raised copper levels.[14]

A woman taking the Pill has blocked off her monthly rhythm of serum copper, and instead retains a permanently high level corresponding to the ninth month of pregnancy. It is normally assumed that this has in itself no effect, being merely a by-product of the raised oestrogen level. However, a survey which looked at three groups of women - one group of pregnant women, one group using the Pill, and a control group which was neither of these - found that the copper serum levels distinguished between these groups better than any other biochemical indicator:

> The most spectacular change noted in this study was in the serum copper levels. There was an almost perfect separation of individual data between the subjects who were not receiving oral contraceptive agents(OCAs) and subjects either using OCAs or who were in their third trimester of pregnancy. One may say that the level of copper in a nonpregnant woman's serum is an almost specific determinant as to whether she is using OCA.[11]

In other words, solely from the serum copper concentration, one could pick out those women who were either pregnant or on OCA.

---

[10] M.Lindler and H.Munro,"Iron & Copper Metabolism During Development,' *Enzyme*,1973,15,pp111-138.
[11] Horwitt et.al.,'Relationship between levels of blood lipids,vitamins C,A and E,serum copper compounds...in women taking oral contraceptive therapy', Amer. Jnl. Clin. Nutrition, April 1975, p406.

The researchers found no other substance in the blood which distinguished these categories so well. This evidence suggests that copper has a dynamic role in the reproductive process, rather than just being a by-product of the raised oestrogen.

We turn next to a quite different mode of contraception, the intra-uterine device or coil. In the early 1970s it was discovered that coil contraceptives using copper were much more successful than previous coil designs. The 'copper-7' coil became the most popular design and was marketed world-wide, used chiefly by women who have already had one child. Despite intensive research however, no-one had any idea as to the mechanism whereby copper in the coil helped prevent conception. Copper ions have a biological action on the inside of the uterus, preventing implantation of the fertilised ovum. Its modus operandi is thus quite unconnected with that of the Pill, where overall blood serum levels are raised.[12] The sole connection is that in both situations a striking Venus-quality is shown by copper's behaviour.[13]

Sweetness of taste has been related to body copper levels: a group of Japanese researchers found that the level at which sugar could be detected in solution depended mainly on the level of copper in the blood serum.[14] It was found that subjects could detect sugar dissolved in water more readily during the afternoon when their copper levels were higher. In all, the function of copper in female physiology surely deserves further study.

Water brass Art

The Venusian nature of copper reveals itself beautifully in the new art form developed by Mr John Byde. I meet up with him at

---

[12] G.Osler and M.Salgo, 'The Copper Intrauterine Device and its Mode of Action',New England Jnl.Med.,1975,292,p433.

[13] Astrologer Maggie Hyde noted this phenomenon. To quote (with permission) from a letter of hers to the author: 'After much effort, I finally confirmed that doctors simply do not know why copper acts as a contraceptive. They simply know that if you put copper into female rabbits, they don't produce little rabbits, and the same happens with women. The theory of correspondences, however, would suggest that if you put VE in the domain of MO, you give ♀ (love) dominance over ☽ (motherhood).'

[14] Hiroshi Torigoe of Tittori University,Japan; an abstract appeared in the Yonago Acta Medica,March 1959; the original in Japanese was published in the journal of the Yonago Medical Association, 1958. (Source: Dennis Elwell)

summer festivals where he gives talks about alternative energy systems. His method, he explained, involves putting a small brass plate into a mountain stream allowing the water to drop over it for several days. There is no false colour or photoshopping in these images here, they have only been magnified. You are maybe looking at a centimetre across in these images. I suggest that any disturbed children could benefit from watching a tranquil sequence of these images http://www.waterbrassart.com/contact/ and that they could inspire children's art.

They have an elemental power and take us away from words for a while. John Boyd and his partner have to 'feel' something about the locations they choose for have some images made during Venus conjunct Jupiter as happens once a year (a good time) versus say Venus square Saturn as a bad time! (Search for video, 'Water brass art montage').

If you need reminding of how copper can produce the lovely rainbows of colour, try waving a sheet of copper over a gas flame, see what is imprinted there. You may wonder, can somebody give you a chemical explanation of this? Well no, I suggest we rather need an *al*chemical explanation, that this is the Venus-being of copper expressing itself.

Fyfe on Copper

A 1978 study by Agnes Fyfe, *'Die Signatur der Venus im Planzenreich'*[15] has an interesting approach, where the height to which a 1% copper acetate solution rose through plant sap extracts on filterpaper was recorded through the course of a year. She found that the height varied with Venus' position in the sky. During a conjunction with the Sun, Venus is invisible, and then grows becomes most prominent in the evening sky when furthest from the Sun (as seen from Earth). Fyfe claimed that copper solutions rose up highest in her filterpapers when Venus appeared highest in the sky as either the Morning or Evening Star, ie when it was furthest from the Sun.

In other experiments Fyfe looked at sap extracts from iris, hellebore and mistletoe on filterpaper pictures using gold, silver and copper solutions, during conjunctions involving Venus and the Sun or

---

[15] Stuttgart 1978 (90 pages), no English translation.

Moon. She claimed that these filterpaper patterns changed in a reproducible manner during these events. In particular, mistletoe was used; the suggestion here is that this parasitic plant, once sacred to the Druids, has no roots in the earth, so can be somehow more responsive to the patterns of the sky.

Fyfe's experiments have been described as difficult to interpret,[16] but she spent many years picking these plants, extracting the sap and making the filterpaper pictures on a daily basis, for medical purposes, and therefore would have become very familiar with any rhythms involved.[17]

The experiments were conducted in a laboratory where the temperature was held at 20°C, and at a relative humidity of 70%, so seasonal fluctuations could be eliminated. Fyfe on the whole avoided numerical evaluation of her experiments, which makes it difficult for others to interpret the many filterpaper pictures shown in her papers. She worked at the Lucas Klinic, a cancer hospital at Arlesheim near Basel, which uses medically prepared extracts of mistletoe.

---

[16] *Recent Advances in Natal Astrology* Dean & Mather, 1977 (published by the Astrological Association) p233: 'Without adequate quantitative data her (ie, Fyfe's) results remain almost impossible to interpret.'

[17] The best summary of her work in English is in John T Burns' online opus *Cosmic Influences on Humans, Animals, and Plants: An Annotated Bibliography*, alluding to eight of her publications.

7. AURORA, GOLDEN GODDESS OF THE DAWN

At dawn or dusk, when the sun briefly shines with the hues of liquid gold, we ponder the words *aurum* (the Latin word for gold) and *aurora*. The French words *l'or* for gold, and *l'aube* for the dawn, are cognate with these, as is the English word 'aura'. 'Au' is the chemical symbol of gold: Rudolf Steiner gave what he called the 'Sun-sound' as Aa-oo, which we express with these two letters, A-U. We here explore the alchemic meaning of this connection, via the mythic image of Aurora. This topic can be helpful in the treatment of depression.

The fiery, solar nature of gold is shown in the way gold is put to the 'proof', melted in a fierce furnace. Then other metals gradually burn up but the gold doesn't, it remains. Even silver is gradually burnt away, as the 'carats' of the gold creep up through the fierce heat. The word 'proof' originally had this meaning, of being put to the test. There was one other metal that would endure the fierce heat and that was platinum, which eventually came to be recognised as the '8th metal' coming in from the New World.

For centuries man has collected the traces of finely-dispersed gold and concentrated it into ingots, to encourage greed and avarice. That was so wrong! The being of gold is experienced best when finely diluted, e.g. in coloured glass, or beaten out thinly into gold foil so it can glitter in the sunshine. Philosophers should study how to make colloidal gold - a colloidal solution of high dilution –where not a lot has been written on the subject since Michael Faraday's essay of 1857, which showed how to make its different hues.[1] Gold needs to be beaten-out and exposed to glitter in the sunlight on sacred temples and palaces of magnificence that are communal, and decorating the beauty of womankind. But under the ground -

---

[1] Michael Faraday *Experimental Relations of gold and other metals to Light* 1857 Philosophical Transactions, 147, Part I, pp.145-181.Somewhat earlier, the astronomer Sir John Herschel experimented with colloidal gold for photography.

'There are however no veins of gold as there are of other metals.'[2]

Why should that be? Gold is found finely irradiated as it were through the Earth's crust, denser in some places than in others. Mother nature doesn't concentrate it and in a sense neither should we. Beaten out to be more delicate than butterfly's wings, it can be used in all sorts of decorative ways. Nanotechnology-type uses of gold appear, in medicine, in semiconductors, fuel cells, etc.

Total Eclipse of the Heart

Depression has overtaken back pain as the number one cause of days off work in today's Britain. I'd like to suggest an *Aurum-Aurora meditation,* maybe with gold taken as a medicine, is here relevant. A homeopathist knows, that if someone has suffered from a 'total eclipse of the heart' say from a relationship going wrong, then a high dilution of gold can be quite an effective medicine: "When a patient reports that he has been depressed all of his life, think of Aurum."[3] "Gold has great remedial virtues, the place of which no other drug can supply" wrote Samuel Hahnemann, the founder of homeopathy. How true!

> And having myself used it in practice for several years, I have come to regard it in the same light: I cannot do without it. To my mind there are varieties of disease that gold, and gold only, will cure, and others that gold, and gold only, will alleviate to the full extent possible, and not a few of these varieties of disease are of the gravest nature. As a heart-remedy alone it claims the earnest attention of every medical man.[4]

Such accounts of the 'Aurum-deficient' person apply, I suggest, to our civilization as a whole:

> By far one of the most powerful anti-depressive remedies is *Aurum metallicum* or the metal gold. Indeed, it does not have any healing powers in its crude form, but prepared according to homeopathic methods it becomes a gem. It

---

[2] C. Budd *Of Wheat and Gold, Thoughts on the Nature and Future of Money* New Economy 1988, p.47.
[3] www.herbs2000.com/homeopathy/aurum_met.htm
[4] Burnett, *J. Gold as a Remedy in Disease* 1879 Finsbury Park, London

covers the classic situation in which one person dies after many years of a happy marriage, and the other partner dies within a few months. The surviving partner will say, 'I lost the sunshine in my life,' meaning they lost all purpose in life. *Aurum* will also help senior citizens in nursing homes where loneliness and a lack of purpose often bring an *'Aurum* state' of emptiness and despair. *Aurum* also covers physical and emotional pain as well as the desire to commit suicide. Aurum is able to alleviate physical as well as emotional pain, as I have seen so many times in my practice.

Responsibility can wear a man thin. After all, he lives in the constant fear that someone may come and rob the treasure which he has the duty to guard *Anxiety; Cares, full of worries; Fear from noise at door.* In a sense he acts like the Minister of Finances who answers to the President and is responsible for keeping the wealth of all people ... No, it is out of duty and a higher sense of responsibility that he guards the treasure ... One day he comes to work and finds that all the gold has been stolen. This is the point where 'the *Aurum* pathology' begins. In keeping with this idea, he develops *Anxiety of conscience as if guilty of a crime.* He feels as if he has let down both the people and the President *Religious despair of salvation.* This emotion is so deep that he sees no way out - he has utterly failed *Sadness with suicidal disposition.* His single symptom *Cheerful while thinking about death* is a beautiful summary of the broken down state: Death is the way out, the only way to purge the soul of the sin.[5]

The gold has vanished from Fort Knox. Where has it gone? Has Fort Knox got five thousand tons of gold, or not? Germany has asked for its large store of gold in the US to be repatriated to Germany – they'll be lucky! Are the ingots remaining just tungsten coated with gold? Senator Ron Paul requested a check on this, in vain.

---

[5] http://www.drluc.com/homeopath-depression.htm

The Goddess Aurora (*Eos* or *Auos* in the Greek)

The dawn-goddess Aurora expresses the cosmic will-to-get-up in the morning. She rises with the first roseate glow of sunrise. We imagine her emerging from the early-morning dawn. She reaches out, her finger touches your heart, and you sense the golden radiance. We speak of the *coronal* arteries around the heart, and the *corona* around the Sun, which have the same derivation. You look into her eyes, and see her innocence!

In mythology her stories shimmered with crimson, yellow, rose and gold. Her appearance in the daily round was ephemeral, which maybe prevented temples from being constructed in her honour – as likewise, she is missed out from modern coffee-table books on goddesses. She complained about this – or so the roman poet Ovid tells us - to Zeus:

> Least I may be of all the goddesses the golden heavens hold (in all the world my shrines are rarest).[6]

To help remedy this sad situation, we here discover twelve *attributes* of this gentle and radiant Dawn-goddess:

> Her crimson doors: "Aurora watchful in the reddening dawn, threw wide her crimson doors and rose-filled halls; the stars took flight, in marshalled order set by Lucifer who left his station last." - Ovid, *Metamorphoses 2.112.*

> Her saffron Robe: "Eos the yellow-robed arose from the river of Okeanos to carry her light to men and to immortals." - Iliad 19.1-2; "Eos appeared in her flowery cloth of gold." - *Odyssey 10.540, etc.*

> Her rosy fingers: Eos (Dawn) comes early, with rosy fingers." - *Odyssey 2.1, etc. (repeated many times).*

> Her steeds of Dawn: "Eos' horses went racing up into the sky today, bearing her all rosy from Okeanos' bed." - Theocritus *Idyll 2.145f;* "Hyperion's daughter [Eos the dawn] expels the stars and lifts her rose lamp on the morning's horses." - Ovid *Fasti 5.159.*

---

[6] Ovid, *Metamorphoses, 13.576.*

Her golden sandals: "Golden-sandaled Auos." - Sappho, *Fragment 103.*

Her golden arms: "golden-armed (khrysopedillos) Auos [eos]" - Lyric IV *Bacchylides Frag 5;* "gold-throned Eos" - *Homeric Hymn to Hermes 326-8.*[7]

Her dewy hair: "Aurora [Eos] arising with dewy hair" - *Metamorphosis 5.446.*

Her cool whip: "So oft hath Tithona [Eos goddess of the dawn] passed by my groans [from lack of sleep], and pitying sprinkled me with her cool whip [the dewy whip with which she chased away the stars]" - *Silvae 5.4.1..*

Her exultant heart: "Eos, heart-exultant in her radiant steeds amidst the bright-haired Horai (hours)" - *Quintus Smyrnaeus 1.48.*

Her healing power: "I long to please Aotis (Dawn-goddess) most of all, for she proved the healer of our sufferings." - Greek Lyric II *Alcman Frag 1.*

Her mortal lovers: "The goddess Eos, who had slept beside Lord Tithonos, was rising now to bring light to immortals and to mortals" - *Oddysey 5.1;* "Eos had just shaken off the wing of carefree sleep and opened the gates of sunrise, leaving the light-bringing couch of Kephalos" - *Dionysiaca 27.1;* "Eos rose from Okeanos and Tithonos' bed, and climbed the steeps of heaven, scattering round flushed flakes of splendour" - *QS 6.1*

---

[7] http://www.mlahanas.de/Greeks/Mythology/Eos.html

Aurora, golden dawn-goddess

> Her dancing-ground: "The island Aiaia which Odysseus reached was 'her dwelling-place and her dancing grounds' and also where the sun arose from. Odysseus and his crew beached their vessel, there upon the sands", (but N.B. she wasn't there when they got there) - *Odyssey 12.1-6.*[8]

In the story, her mortal lovers would get left behind when she had to get up out of bed and open the shutters of dawn, etc. Or if they were lucky, they could enter her chariot, as her steeds Lampos and Phaithon champed impatiently, in Dawn's early glow, waiting to fly in the sky.

We expect a solar goddess to have some twelvefold attribute, and Aurora was connected with the twelve hours of the day:

> Twelve maidens [the Horai] shining-tressed attended her, the wardens of the high paths of the sun for ever circling, wardens of the night and dawn.[9]

In an 'Aurora' workshop one would intone the sound 'Aurora', as well as that of her island *Aiaia* which was her dancing-ground and also where the Sun rose! One could also appreciate the solar themes in songs, 'Total Eclipse of the Heart' and 'Holding out for a Hero' by Bonny Tyler (1983, '84) The latter concerns that liminal realm at the

---

[8] The crew eagerly awaited her: "we beached our vessel upon the sands and disembarked upon the sea-shore; there we fell fast asleep, awaiting ethereal Eos" - but alas she wasn't there.
[9] Quintus Smyrnaeus, Fall of Troy http://www.theoi.com/Titan/Horai.html

time of waking up, before the dreams of the night-time have fled. Such a workshop should really be up to greet the dawn ...

Figure: The mystical masterpiece *Aurora,* painted by Otto Runge in 1808.

Aurora/Eos also has her son slain, in the Trojan war, then she becomes too depressed to get up in the morning etc. This story could also be used in treating cases of depression.

Tchaikovsky's ballet *Sleeping Beauty* was about the Princess Aurora. It involved a baptism-party, where the king had just twelve *golden plates*, so the 13th lunar-unlucky 'wise woman' in the kingdom couldn't be invited. A curse was laid, whereby the Princess would have to bleed when she reached her 15th year.[10] Tchaikovsky shifted to her 18th year, which rather lost the heart and soul of the

---

[10] Jacob and Wilhelm Grimm complete Fairy Tales, 1972, *Briar Rose*, p.203.

fairy-tale: the struggle between solar and lunar motifs, the deep, blood-meaning of the story, went out of the window and instead the theme becomes merely, will the prince arrive and kiss Aurora? I guess that's showbiz - but, he did use the name Aurora, for his doomed solar heroine, for which we should be grateful.

Lion-Heart

If the classical Aurora myth is a bit faint maybe it is because horses are not and cannot ever be 'solar.' They are rather associated with intelligence and service: "The tygers of wrath are wiser than the horses of instruction" (William Blake), and lack any solar quality. Whereas, the white lions now being re-introduced into South Africa[11] do have that majestic solar principle, concerning the affirmation of life - that actually, you do want to get up in the morning. Africa has deep solar themes of gold, lions and the rhythm of the heart in its music. Let's have some solar myths with more of an Afro- slant to them.

Linda Tucker had a city job in London, which involved some beauty-modelling, when she felt the challenge to go to South Africa and help the re-breeding of White Lions into the wild, as there were none left. She had the courage to say 'I will,' and she got up and did it.

---

[11] http://www.whitelionshomeland.org/ *Mystery of the White Lions, Children of the Sun-god*, Linda Tucker, 2003.

A small number of these re-bred white lions have now been re-introduced back into the wild.

Her book, *Mystery of the white Lions, Children of the Sun-God* explores African traditions whereby lions were known as the guardians of gold and gold was 'the lion of metals.'. We feel the solar being-ness of the lion, which is something without words, but these words may help: *radiance, splendour, majesty, heart-power.*

She sought for the solar myths of Africa, and the special role of the majestic white lions. She found that: 'the maps of the goldfields drawn up by researchers at the turn of the 20th century detail how the smelting sites are also clustered around the 31º east longitudal line.' Her astonishing conclusion was that: 'On the African continent, the distribution of gold is concentrated around this meridian, with the gold yield in and around this area in the southern hemisphere recently accounting for over half the world's gold production.' (p.176) She believed this was a 'gold artery running through the entire continent of Africa,' which was somehow associated with the Nile. That river 'followed this meridian to its delta in the north,' and this 'Nilotic Meridian ... defines the centre of the earth's landmasses.' This meridian connects with the pyramids at Giza, and is the longitude line which covers more dry land than any other. The white lions are indigenous to Timbavai, their homeland, which is on that meridian line, way down south of the Sphinx.

Magic of Gold

The magic of gold has a connection with the human heart, and relates to that which is everlasting. The Roman natural philosopher-scientist Pliny wrote of gold's healing power:

> Gold is efficacious as a remedy in many ways, being applied to wounded persons and to infants, to render any malpractices or sorcery comparatively innocuous that may be directed against them.[12]

The Bible begins with a garden and ends with a city. Its final, transcendent vision is of a city, glowing golden from within. No Sun

---

[12] Pliny, *HistoriaNatura*, Lib. XXXiii, Cap XXV.

shines above it,[13] and instead the streets have a golden radiance. To quote the late British philosopher John Michell, "this wonderful vision of a glittering, translucent city is the climax of St John's Revelation."[14] The numbers involved in the dimensions of St John's heavenly city are solar: 12 and 144.

If we may be allowed to consult the *Satanic bible* of Anton Lavay (California, 1969), it gave this advice to aspirants:

> Ideally the chalice should be made of silver, but if a silver chalice cannot be obtained, one made from another metal, glass, or crockery may be used - anything but gold. Gold has always been associated with white-light religions.[15]

Satanists mustn't use gold - it's too heavenly!

If we imagine a temple to Aurora, one of its East-facing windows would be coated with a monomolecular layer of gold, which makes it green as light shines through it; while others glow pink with the hues of colloidal gold. On the other side, a window would show images of present solar flares, with close-up pictures of the grand displays and storms on the surface of the Sun.

Golden Glass

"And the city was pure gold, like unto clear glass"
Book of Revelation, XXI, 18.

Glassmakers use high-temperature ovens to work with gold, that can hold around six hundred degrees centigrade. A gold compound is added into the molten glass, and it's then cooled, then slowly re-heated and cooled again until it starts to glow pink as the colloid mysteriously forms. It can produce a lovely pink at about ten parts per million, and ruby red at a slightly higher concentration, a tricky and skilful art.

The secret of making red glass was found in Bohemia in the 17th century.[16] This was a blow to the pride and prominence of

---

[13] "And the city had no need of the Sun, neither of the Moon ..." Rev. XXI 23.
[14] John Michell, *The Dimensions of Paradise*, 1988, p.22.
[15] Anton LeVey The Satanic Bible, 1969, p137.
[16] "The ideal ruby color is attained when many small gold particles are formed, being neither too small, nor too large. The evidence is in the color. A very small number and size of gold particles will color the glass firstly a pale brown and then with time,

Venetian glass-makers, who had tried unsuccessfully for years to make it. Johann Kunckel re-discovered how to make gold ruby glass, in Brandenburg, Bohemia around 1670 ... Another sometimes named as the inventor of gold ruby glass is Andreas Cassius, whose purple-red pigment called 'Purple of Cassius' was hard to make and sometimes used to colour glass red.

The film, *Heart of Glass* (Herz aus Glas) made by Werner Herzog in 1976 is set in 18th-century Bavaria, It features a small village renowned for its "Ruby Glass" glass blowing works, where the foreman of the works dies suddenly without revealing his secret of the Ruby Glass: the town slides into a deep depression. The owner of the glassworks becomes obsessed with the lost secret. (available online)

Other chemicals produce red glass, but none have the special magic of gold ruby:[17] The beauty alone of gold-ruby glass justified neither the tremendous efforts made in its development nor the high prices which these glasses brought. It was no doubt the mystic power attributed to gold and the ruby colour produced by it which was responsible for the extraordinary demand.[18]

Where did the red come from? Traces of tin or lead seemed to be necessary, "their presence more or less controls the development of the colour," or else a mere pale pink hue would result. The ruby hue is spoilt if the glass is allowed to cool slowly, it must be fairly rapidly cooled, before re-heating when the nuclei of gold ground with the glass to make the hue. The 'thermal history' of the glass is important. At least in theory, "appropriate treatment of a gold-containing glass could lead to practically all colours of the rainbow."[19]

It's worth a visit to the British Museum to see the ancient-Roman Lycurgus cup, featuring the story of Dionysus: it has a green colour

---

the brown gives way to a purple shade. Continued heating slowly increases the size and number until a ruby eventuates. If the Gold Amethyst is arrested at the purple stage the color may be too deep a purple. Longer firing will lighten the hue." http://gafferglass.com/technical/instructions_main.htm

[17] http://www.glass.co.nz/gibruby.htm *The Art of Glass* by Antonio Neri 1662 is credited with having the first description of making gold-ruby glass written in plain language: Muriel West, *Ambix* 1961, IX p.113.

[18] W.Weyl, *Coloured Glasses* 1951 Sheffield, p.380.

[19] Ibid p.382. For tin found in old 'ruby glass' together with gold, see Catherine Louis and Olivier Pluchery *Gold Nanoparticles for physics, chemistry ad Biology* 2012, p.12.

except when illuminated from within when it glows red. Someone had the knack of putting minute amounts of gold and silver nanoparticles into its glass, creating this optical effect, but that art was then forgotten.

These days, nanotechology institutes are springing up: tiny gold nanoparticles can now be produced as Platonic solids – the regular polyhedrons tetrahedra, hexahedra, octahedra, icosahedra, and doecahedra. The story of making gold glass is now being re-told in terms of how such nanoparticles form within the glass, which produce the different hues. A new interest in gold has come about through the small nano-structures made from gold atoms, used for all sorts of catalytic processes.

Purple Gold

Chemistry nowadays has vanished from our civilisation as an experiential science. Let's try to bring it back again, to regain the wonder and mystery that should belong to this subject. To see the lovely hues of gold, prepare a dilute solution of gold chloride, say 0.01%, or maybe some higher dilution, and a solution of Rochelle salt (a weak reducing agent), say 1%. Add the one to the other, and if you are lucky the solution will slowly grow into a lovely purple. I once found the solution went black – maybe it was too concentrated, or not clean enough. Then by the next day this had turned into a lovely ruby red, just like the filterpaper-picture shown above (by Kolisko, using 1% gold solution in the sunshine, 1936). Put these solutions on your window-sill and admire them, they will last a week or so.[20]

What used to be called 'purple of Cassius' is made using a tin solution. Add stannous chloride solution to gold to make it. I didn't succeed, and instead obtained a rich hue of golden-amber. Again, why not put this into a phial on your window-sill. As the Sun-metal, it needs the sunshine.

---

[20] One can use formaldehyde as the reducing agent, but it doesn't give such a nice purple. One first makes the gold solution alkaline by adding, say, some bicarbonate of soda, and then add a drop of formaldehyde. Nothing happens for quite a while, and then slowly the solution grows purple. This is a rather muddy hue, not so attractive as that produced by the Rochelle salt. (See note 3 Chapter 3) See my attempts here: www.sciencegroup.org.uk/kolisko/gold.htm
www.sciencegroup.org.uk/kolisko/

Mrs Kolisko obtained striking colour-patterns on filterpaper by rising solutions of gold in sunlight (front cover image). The sunlight itself reduces the gold. She published her gold-experiment over the 1961 total eclipse of the Sun, claiming that the gold image varied over the course of the eclipse. I found it difficult to repeat her gold experiments, the whole effect is so temperamental.[23] Using a 1% solution of gold chloride, one can allow it to rise or spread out in sunlight up a filterpaper. When its dry, do it again, so that it rises somewhat higher.

Astronomy puzzle

The Cup dipped into the sun. It scooped up a bit of the flesh of God, the blood of the universe, the blazing thought
  Ray Bradbury, the Golden apples of the Sun

Astronomers have a theory which explains what makes the sun shine, but this may not be working too well[21]. It encounters some problems:
* *If* Sol is powered by thermonuclear heat at its centre – as Fred Hoyle assured us – how come no-one has been able to make stable thermonuclear power (i.e., fusion energy) happen here on Earth? Over half a century, massive funding of such projects has got nowhere. Such fusion reactions (hydrogen-helium) depend greatly upon temperature, and produce much heat, so inevitably they are going to go bang.
* The Sun's 11-year cycle is *not* predicted by any nuclear-fusion theory. Everything on its surface varies with that 11 or rather 22- year cycle, which should not exist, according to any nuclear-fusion model of solar activity: yet it does and appears very stable; it is found in old records of sedimentation etc from millennia ago.
* The temperature-gradient on the Sun's surface is the wrong way round: the corona of the Sun glows at millions if degrees centigrade, while the surface photosphere is cooler at around six thousand degrees, and below that, which we only get to see inside a sunspot, its around four thousand degrees. Sunspots have a dark interior *because*

---

[21] See N.K., 'Power of the Sun', *Journal for Star-Wisdom,* 2015, Ed. Powell, pp.83-6 (online).

they are cooler. If heat is coming out from the sun's centre, that temperature gradient should be the other way round.

* The Sun rotates differentially, much faster at is equator than at is poles. That should not be happening, if heat is diffusing out by convection currents from the centre as astronomers tell us, because those currents ought to have slowed down any such differential rotation.

The basic theory of what makes the sun shine is quite shaky and unverified because no-one has ever seen a stable nuclear fusion reaction, and there are theoretical reasons for doubting whether it is possible. Nothing on the sun's surface looks as if it's produced by heat diffusing out from the centre – it all *looks* electrical-magnetic in its nature.[22] The solar corona should not exist on the standard model. It's hard to think of any scientific theory which so greatly fails to explain just about everything which requires explaining, as does the current model used by astronomers for the Sun.(24) Of the three different layers of the sun's surface, the photosphere, the chromosphere and the corona, each behave in quite different ways, and need different types of light to photograph them (visible light, UV light and X-rays respectively). We may gain the impression of a living organism here, which can somehow stay hot and stable over vast aeons of time.

The heart-beat of the Sun pulses every twenty-two years. The upward climb of its new solar cycle was quite delayed, and its present (2014) peak is rather faint. Its entire magnetic field reverses North-to-South every eleven years. We may see this as somewhat biological, a *cardiac* process of systole-diastole. Astronomers will not want to comprehend the *meaning* of the wonderful stuff they discover. They've managed to get SOHO sun-satellites right outside the plane of the solar system to detect this stupendous process. Jesus said, "As a man thinketh with his heart, so is he" and maybe if astronomers could think with their hearts a bit more we might get somewhere.

From an alchemic perspective, Sol is unique. No other star is known to have a self-reversing magnetic field. It's not 'just an ordinary yellow dwarf' as astronomers are prone to remark. It is the fiery, pulsating heart-centre of our solar system.

---

[22] Two excellent books are, *The Electric Sky, a challenge to the myths of modern astronomy* by Donald Scott 2006, and the *Electric Universe, A new view of the Earth, the Sun and the heavens* by Wallace Thornhill and David Talbott, 2007.

We conclude with the archangel Raphael's words in Goethe's *Faust*, in the Prologue in Heaven:

> *The Sun-orb sings, in emulation,*
> *'Mid brother-spheres, his ancient round:*
> *His path predestined through Creation*
> *He ends with step of thunder-sound.*
> *The angels from his visage splendid*
> *Draw power, whose measure none can say;*
> *Thy lofty works, uncomprehended,*
> *Are bright as on the earliest day.*

Suggested music here: Pink Floyd, *Setting the dials for the Heart of the Sun,* 1968; *Roar,* Cathy Perry, 2013.

8. CHEMICAL HISTORY AND ALCHEMICAL MYTH

In mythology, the Age of Gold came first, followed by the Age of Silver. Originally only the metals gold, silver and copper were known, as these occur naturally. However, silver and copper only occur in the native metallic form in very small quantities, and so gold would have been the commonest metal in some countries. Silver ores tend to occur together with lead and so lead smelting would have produced silver.

The metal smelted over the longest period of time is not copper, as commonly supposed, but lead.[1] Lead's low melting point made it easy to extract. Its commonest ore was galena, lead sulphide, a black, shiny substance which can be reduced to the metal at fairly low temperatures. The earliest products made of lead date from the 7th millennium BC.

In the 4th millennium BC, objects made from copper begin to occur. Decorative objects, bowls and amulets were made from it. It is presumed that these would have been made from native copper. A thousand years later, ores of copper begin to be smelted -- malachite, azurite and chalcopyrite. A relatively high temperature, 1200 degrees, is needed for this, such as would not have been reached by a camp fire.

The Bronze Age begins at this time. Tin was discovered in about 3000 BC and became used mainly for bronze making, from which tools and weapons could be constructed. Hesiod in the 7th century BC told of how, in the Age of Bronze:

> The weapons of these men were bronze, of bronze their houses, and they worked as bronze-smiths. There was not yet any black iron.

---

[1] Gale & Gale, 'Lead and Silver in the Ancient Aegean', *Scientific American*, June 1981, p.142.

The first records of metal-planet correspondences date from 2000 BC or earlier.[2] The Assyrians and Babylonians named the metal iron after Ninip their god of war, and named lead Anu after a sky god with some resemblance to Saturn. Tin was obtained from the British Isles, which the Romans called the Cassiterites because of this, and passing along the great trade routes received the name of a north Italian (Etruscan) version of Jupiter, Tinia.

Gradually, the category of metals came to be viewed as sevenfold. Aristotle mentions all seven of the traditional metals, referring to 'liquid silver' which must mean mercury.[3] This does not necessarily mean that he believed in just seven metals - he could well have regarded brass, electrum and steel as other metals. The distinction between an alloy and a metal was not always evident.

As mercury became generally known and available in the early centuries A.D., the correspondences come to resemble the traditional ordering here used. Very little has been written on the history of this subject. E.J.Holmyard claimed that the linking of seven metals with seven planets was adopted by the Greeks of around the fourth century B.C.[4] Although he was quite an expert on these matters, I have failed to discover such an antiquity for the doctrine. Lists of sevenfold metal-planet correspondences appear from the second century AD onwards, and may have derived from the mysteries of Mithras

Seven Steps to Heaven

The earliest tradition of 'seven metals' as linked to the spheres of the planets occurs in the religious system of Mithraism as described by Celsus, who was writing in Rome about A.D.180. Unfortunately, the

---

[2] Partington, J.R., 'The Origins of the Planetary Symbolism of the Metals', *Ambix*, Jnl. for the History of Early Chemistry, Vol. 2. p61.

[3] Partington, J.R., *A History of Chemistry*, Vol. 1, 1970, p.101. This volume (posthumous) is much recommended for its treatment of Hermetic and alchemical beliefs of antiquity. His published three History of Chemistry volumes are works of staggering tedium, each over a thousand pages long, however this one (which evasively is just entitled 'Theoretical Background') surely remains the finest history of ancient alchemy/chemistry with its Hermetic-Neoplatonic background. Each page has about ten references, and Holmyard was readily able to quote Greek and Latin texts: such a memory will not recur!

[4] Holmyard, E.J. *Alchemy* 1957, p.21.

sole remaining text to describe this belief is by an opponent, as Celsus' text is lost. Origen's *Contra Celsum* of A.D. 248 said:

> These things are hinted at in the accounts of the Persians and especially the mysteries of Mithras which are celebrated amongst them...There is a ladder (or stair) with seven gates and above that an eighth gate. The first gate is of lead, the second tin, the third of bronze (or copper), the fourth of iron, the fifth of mixed metal, the sixth of silver, and the seventh of gold. The first gate they assign to Kronos, indicating by lead the slowness of this star...

The Greek phrase for 'mixed metal' may have referred to electrum, a gold-silver alloy used for coinage. A mixed-up list of correspondences ensues, such as 'the fourth to Hermes, since both Hermes and iron are fit to endure all things and are money making and laborious,' a doubtful-sounding analogy not found in other lists by alchemists of the time and probably inserted to denigrate the whole philosophy. The eighth gate corresponds it has been supposed to the sphere of the fixed stars. The steps of a ladder of Mithras would not have been constructed of the actual metals, as gold and silver would have been too expensive, but were painted to represent them. The quote by Origen continues,

> He (Celsus) next proceeds to examine the reason for the stars being arranged in this order, which is symbolised by the different kinds of matter; musical reasons, moreover, are added or quoted by the Persian theology, and to these again he tries to add a second explanation, also connected with musical considerations.[15]

And that is all! One surmises that 'to examine the reason for the stars being arranged in this order' would indeed have been difficult, because the order given was not astronomical, but alchemical: a scale of perfection from lead the 'basest' metal up to incorruptible gold. Later alchemists would put copper above iron in this list, but as iron was then a relatively new metal, presumably it was then regarded as more valuable than copper. With that slight adjustment, and with the insertion of quicksilver in place of 'mixed metal', this list remained for the best part of two millennia the standard ordering used by alchemists

---

[5] Partington, J.R., op. cit., p.299

as their 'scale of perfection', as part of their alluring notion that the baser metals could be transformed into gold.

Aphrodite's Isle

The wealth and sea-trading power of Cyprus arose from the discovery of copper mines in the 3rd millennium BC. It became the main source of copper for the ancient world, and the mines are still in use today.

The origin of the word copper is intertwined with the name of Cyprus. Experts are unsure whether the island gave the metal its name, or vice versa. Cypress trees may have been named from it (*Cupressus*, latin). Or perhaps the island was named after the cypress trees? No one is sure. One thing is for sure, the cypress trees were cut down to smelt the copper. Antiquity regarded Cyprus as a special place, and traditions and legends have always viewed it as "the Sweet Isle"':

> *O to flee hence unto where Aphrodite*
> *doth in Cyprus, the paradise-island, dwell,*
> *the sea-ringed haunt of the love-gods mighty*
> *to weave the soul-enchanting spell,*
> *or the fields where untold is the harvest's gold.*
>
> Euripides, *Bacchae*[6]

Enigmas of Mercury

Cinnabar, the bright red sulphide ore of mercury, was well known as a pigment in antiquity. But it wasn't easy to reduce the ore to quicksilver, because however it was done heat was necessary, and this would vapourise the mercury. Many an alchemist must have been poisoned in this way. The first record of metallic mercury's preparation comes oddly enough from various Egyptian tombs of about 1500 BC. Small ampoules of the liquid metal were found therein, as gifts for Thoth-Hermes it appears, guide to the soul of the dead.

---

[6] Translation from P.Stavrou, *Cyprus, The Sweet Land*, Nicosia 1971,p.18.

A recipe for fixing this elusive metal was first given by Dioscorides (c. AD 50). Mercury Cinnabar was heated with a reducing agent and the vapour was condensed. The cinnabar was heated on an iron saucer contained in a pot, so that iron reduced the cinnabar, and the mercury was collected off the top of the pot (Iron is the only one of the seven metals which resists amalgamation with mercury).

Pliny the Roman naturalist described a different method of making mercury. Cinnabar was ground with vinegar in a copper pestle and mortar, when shining globules of the liquid metal appeared. The vinegar turned cinnabar into a soluble salt (mercury acetate) and the copper then reduces the mercury:

$$HgAc + Cu^+ = CuAc + Hg^+$$

('Ac' = acetate)

As this was a different method, Pliny assumed it must be a different kind of mercury, so he called it 'hydrargyrun ' (water silver), and the chemical symbol for mercury 'Hg' recalls this name.

Through such transformations the first glimmerings of the idea of a chemical reaction must have stirred in the experimenters' minds. Sulphur and mercury were both given a central place in alchemic theory, both being regarded as elementary - as the metals, apart from mercury, were not. In particular, reactions with mercury were *reversible*. From sulphur and mercury, cinnabar could be formed by heating them together. Thus it was shown that cinnabar was composed of two other substances.

Also, certain salts of mercury would *sublime*. The chloride ($HgCl_2$) was purified by heating in a retort, which would cause long, needle-shaped crystals of the white chloride to grow down from the top of the retort. No wonder Hermes was regarded as mutable and chameleon-like, if his metal was so expert at transformation. The metal mercury appeared to alchemists as a vital key to their art. As Hermes was reputed to have revealed the principles of alchemy on his Emerald Tablets, so did his metal quicksilver seem to show the principles of chemical transformation. An ancient recipe describes the first apparently reversible reaction:

> Now the substance of cinnabar is such that the more it is heated, the more exquisite are its sublimations. Cinnabar will become mercury, and passing through a series of other sublimations, it is again into cinnabar, and thus it enables man to

enjoy eternal life.[7]

This was indeed the classic recipe whereby alchemists impressed their clients. After obtaining mercury from cinnabar, gradual heating of the metal then produces a red calx - but this is in fact the oxide, not any longer the sulphide. And it is the reaction with the oxide which is properly reversible:

$$2Hg + O_2 = 2HgO$$

Joseph Priestley used this property of mercury in the eighteenth century to prepare oxygen. Mercury oxide enjoyed a remarkable medical career as the staple remedy for syphilis, from the sixteenth century when the disease appeared and Paracelsus prescribed the remedy, up until the twentieth century when it was replaced by penicillin. The main practical use for mercury has always been in the extraction of gold and silver from rocks. This so-called amalgamation process was used in the Middle Ages, and is still in use. Auriferous rocks were washed with mercury, which would steal the gold from them and the mercury was then distilled to leave behind the gold.

The Iron Age

In antiquity, iron was an exotic sky-metal, only found in meteorites, and more precious than gold. The Egyptians traded iron from the Hittites for several times its weight in gold. Homer describes how a lump of meteoric iron was awarded as a prize in the Olympic games.

The secret of how to extract iron from its ores, using charcoal and a high temperature furnace in which air is sucked through, may first have been discovered by the Hittites of Anatolia, now central Turkey. They were a fiercely martial race. The technique of making iron was a sec which they kept to themselves. It became generally known over the 5th - 3rd centuries BC in the Roman world, giving to Rome its backbone of Martial power: its steely strength of sword, well-nigh invincible Although iron would not be melted at any temperatures then available it could be forged.

The early making of steel would have been a chance affair. Iron is given strength by the earth-element carbon. Maximum strength is attained when a few per cent of carbon is mixed with the iron: a little

---

[7] Quoted in J.Bronowski, *The Ascent of Man*, 1976, p.123.

too much and it becomes brittle, a sword will snap.

Later on the Vikings developed an improved sword-making technique. Blades were built from many thin strips of iron, each hardened on their surface with carbon. They would be heated and hammered together, and the process repeated several times. Such swords were longer and stronger than those employed by the warriors of ancient Rome; against which old weapons of bronze were of little avail.

In Europe over a thousand furnaces were developed hot enough to melt iron, called blast furnaces. They were tall, and used bellows to pump the air in at the base of the furnace. Henceforth iron could be cast as well forged. From this time onwards the amount of iron manufactured is greatly multiplied, ushering in the modem, magic-flee Age of Iron.

Rulership Patterns

In AD 160 the Syrian astrologer Vettius Valens gave the following correspondences: lead/Saturn, tin/Jupiter, iron/Mars, gold/Sun, copper/Venus, elektron (a gold-silver alloy)/Mercury, and silver/Moon. Variations on the theme appear except for gold, silver and lead over which there is no disagreement. The correspondences stabilise into their present form around the seventh century AD, by which time agreement has been reached as to which are the true metals and which are not (mercury, the liquid metal, being the most disputed). Subsequently the names of the planets are often used for the metals in alchemic and medical treatises, and so taken for granted is the belief that explanations are not regarded as necessary.

These planetary correspondences remain stable amongst the chemists and alchemists of Europe for at least one millenium, from the seventh to the seventeenth centuries. Prior to the seventh century one finds some differences in the lists of correspondences, such as the omission of quicksilver and an alloy in its place. The alchemists and metallurgists of old viewed the metals not as elements but rather as intervals along a scale, a sevenfold scale.

In the year 1029, an Arabian textbook on astrology was published by Al-Biruni. His correspondences were a little muddled, with copper assigned to Mars! These were his mineral rulerships:

Saturn: Litharge, iron slag, hard stones, lead.

Jupiter: Marcasite, tutty, sulphur, red arsenick, all white and yellow stones, tin, white lead, fine brass, diamond.

Mars: Magnetic iron, cinnabar, rouge and mosaics, iron and copper.

Sun: Jacinths, lapis lazuli, yellow sulphur, gold and whatever is coined therefrom for kings.

Venus: Magnesia and antimony, silver and gold and jewels set in these, household vessels made in gold, silver and brass, pearls, emeralds, shells.

Mercury: Arsenic, amber, all yellow and green stones, old gold and quicksilver, turquoise.

Moon: Glass, white stones, emerald, moonstone, silver and things manufactured of silver.[8] The planetary symbols came to be used routinely to denote the seven metals, rather as chemists today use the chemical symbols from the latin for the elements. They were used not only by alchemists, but by metallurgists and pharmacists or indeed anyone concerned with chemistry.

Newton's Notebook

Isaac Newton left behind reams of alchemic notes, covering three decades of his life, peppered with the 'astrological' planetary glyphs. I suspect this is what has led astrologers to their mistaken idea that Newton believed in astrology. It was usual for alchemic and astrological beliefs to go hand in hand, but Newton's belief in alchemy was not associated with any degree of belief in astrology, as far an anyone can tell.[9] In the records of his chemical experiments, through the latter half of the seventeenth century, what we nowadays think of as

---

[8] Al-Biruni, 'The Book of Instruction in...Astrology', Ghaznah 1029, translated R.R.Wright, London 1934. See also Lee Lehman's *The Book of Rulership*: 'al-Biruni represents Arabic astrology from the Mediaeval period' (p.2).

[9] Dobbs, B.J. *The Foundations of Newton's Alchemy* 1975. See my: 'Newton, Isaac' in the *Biographical Encyclopedia of Astronomers* Volume 2, Ed. Hockey, 2007 (http://dioi.org/kn/newton-bio.pdf).

the planetary glyphs were used throughout to symbolise the seven metals: ☉ for gold, ♀ for copper, ♂ for iron and so forth.

The number of different metals was then accepted as being seven, as had been the case for many centuries. Like the seven strings of Apollo's lyre, each was invested with a distinct character and identity. Drawing on the widespread literature on alchemy then available, Isaac Newton summarised these correspondences as follows:

Planetae septem

Noah	Ham	Chus	Assur	Thoth	Pathros	
Kiun	Hammon	Hercules	Astarte	Anubis	Orus	Bubash
Saturn	Jupiter	Mars	Venus	Merc.	Apollo	Diana
♄	♃	♂	♀	☿	☉	☽
Plumb.	Stannum	Ferrum	Cuprum	Argenvive	Aurum	Argent.

This was done in the 1680s, as part of a work he never got round to publishing. The list contains – as a biographer of Newton explained – a metal, a planetary glyph, Egyptian and Roman deities and 'a member

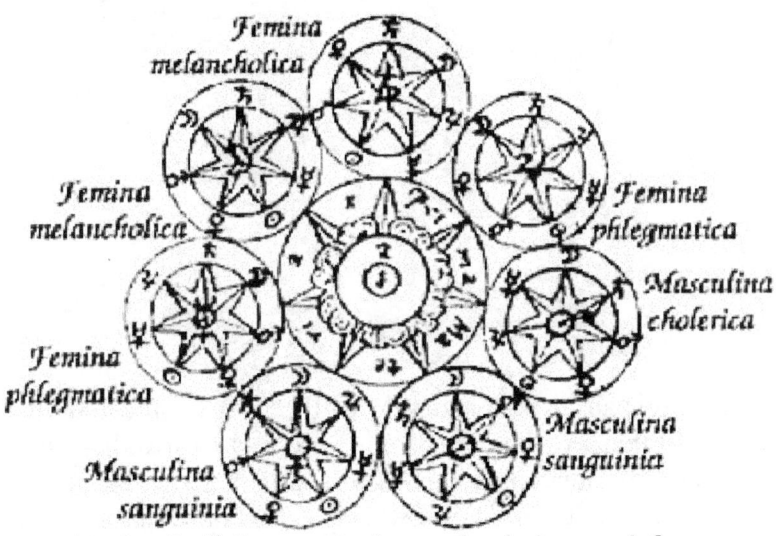

of Noah's family.'[10] ('argentvive' was the latin word for mercury, it meant 'living silver')

Newton devised (but didn't publish) a complicated alchemical mandala, and entitled it: *Lapis Philosophicus cum suis rotis*

---

[10] Westfall, R.S. *Never at Rest, a biography of Isaac Newton*, 1980, p.359.

elementaribus' here shown (translation, 'The Philosopher's Stone with its elemental wheels,' or 'rotating elements'). He put the seven planets/metals in their Ptolemaic order from Saturn to the Moon around a circle (See Chapter 2, tables 1 and 2). Each of the smaller circles then has a sevenfold ordering of the planetary glyphs. Was this arrangement alchemical, astronomical or astrological?

Next to this diagram he wrote out instructions as to what colours hade to be used. Here is a modern reconstruction of this four-temperament seven-planet mandala.[11]

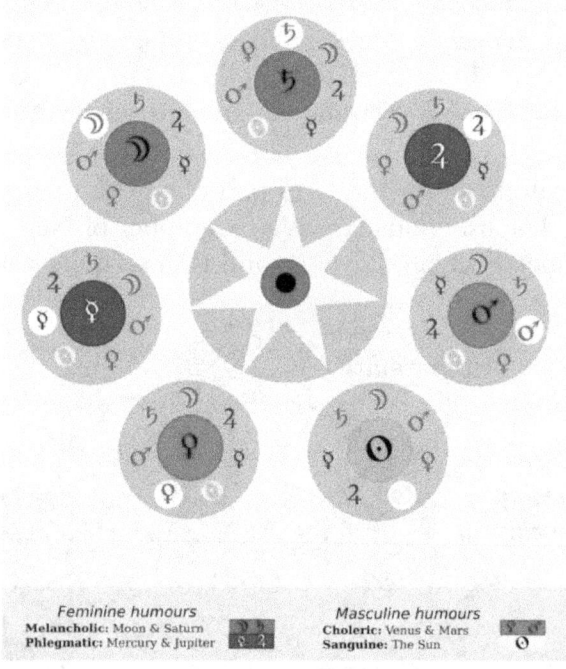

Planets are paired off each side of the Sun (in the old Ptolemaic sequence) so that as Mars and Venus are locked together, so too are Mercury and Jupiter, plus the Moon and Saturn. These pairs are always together in each of his wheels. He links 'humours' to each planet, eg 'Femina melancholia' for the Saturn-wheel at the top of his figure. He's trying to reconcile a sevenfold scheme with the old four temperaments, such as phlegmatic and melancholic.

---

[11] Thanks to Derek Norcott for his reconstruction of Newton's mandala

Traditionally, linking planets with the temperaments in this way would be called astrology. I guess this is the nearest Sir Isaac got to that subject, which he regarded as 'unlawful.'

Newton's comment on Mercury was quoted at the start of Chapter five, published in his 'Queries,' attached to his *Optics* publication. It is worth meditating on this text: it is strictly a list of chemical facts - and yet, *Mercurius* is assuredly still there, is he not? Hermes-Mercury as the guiding spirit of alchemy, elusive trickster and master of transformations is very much still present – as he would be totally absent from all chemistry textbooks, in subsequent centuries.

Newton believed in the special significance of the number seven. He inserted a seventh colour into his spectrum of colours which no-one could see, indigo, in accord with his theory that they were related to the seven notes of the musical scale. His *Optics* was composed in seven sections, his description of the diffraction rings seen around lenses were sevenfold, his lunar theory had seven stages to it, and so on. He appears as a pivotal figure, still using the seven planetary glyphs, but standing rather at the terminus of the alchemic tradition.

Isaac Newton and Robert Boyle were both members of Britain's newly-formed Royal Society. Boyle published a note in the Royal Society's journal, reckoning he might have made the legendary 'philosopher's mercury' and how medicines made from this could be most effective – however, should this special mercury 'fall into ill hands' there could be a 'political inconvenience'. He invited any advice on this topic, to whether he should keep quiet about it or give the recipe!

This was a time when the cognoscenti in Europe such as Robert Boyle and Leibniz were inclined to believe stories indicating that alchemists could indeed transmute metals (see next chapter). Boyle believed that one needed some special kind of Mercury to do this. What was Newton's view?

The young Newton – who had only just joined the Royal Society a few years ago, as a result of his prism experiment with colours and making a telescope - decided to reply to this request .He wrote enigmatically to Robert Boyle, 'If those great pretenders bragg not' – alluding to the alleged sages who could make gold! They surely had the secret of mercury:

> there being other things beside ye transmutation of metalls (if those great pretenders bragg not) wch none but they understand.

Newton was advising Boyle to *keep quiet* about whatever it was that he knew! This was in the year 1676 when Newton was in his thirties, and his hair had recently gone grey – probably from all of his experiments with mercury, or so he remarked, maybe in jest. Urging secrecy, he said that Boyle's special mercury 'may possibly be an inlet to something more noble, and not to be communicated without immense dammage to ye world.'[12]

Newton was not quite sure whether Hermetic philosophers existed – but, if they did, they knew a thing or two:

> ...a true Hermetic Philosopher, whose judgement (if there be any such) would be more to be regarded in this point then that of all the world beside to the contrary, there being other things beside the transmutation of metals (if those great pretenders bragg not) which none but they understand.[13]

Who else gets praise like that from Newton? He is saying that there might be big secrets concerning mercury, but if so Robert Boyle should keep quiet about it! The new science aimed to share matters in their weekly meetings and to publish them in their journal, not keep them secret as had the Mediaeval guilds – but, this was an exception!

That was a 17th-century view, whereas in the 18th century Newton like the rest of the world became a deal more sceptical. Experts remain unsure just how much he believed these matters. He wrote reams on alchemy, but published almost none of it. One can say that the language of alchemy was a matter of urgent importance for the creative decades of Newton's life but that it gradually faded out after his nervous breakdown of 1693, when he became a public figure involved with London's Mint and Royal Society. Well over a hundred volumes on alchemy were found in his library after his death.

---

[12] P.A.Fanning *Isaac Newton and the Transmutation of Alchemy*, 2009, p.162.
[13] Newton's letter to Oldenburg (secretary of the Royal Society, who would have sent it onto Robert Boyle) 26 April 1676
http://www.newtonproject.sussex.ac.uk/view/texts/normalized/NATP00268

I here can't resist adding, that his nervous breakdown which lasted about a week was timed exactly by Neptune going over his natal Mercury.

Robert Boyle, known as the 'father of modern chemistry,' testified before Parliament that he had witnessed the transmutation of base metals into gold – he was asking them to repeal a three-centuries old act banning alchemic transmutation: 'none from henceforth shall use to multiplie gold and silver' – the act passed by Henry IV in 1403. The act repealing this was passed in 1688, in the first legislative session of William and Mary after the so-called 'Glorious Revolution:' 'An act to repeal the statute made in the fifth year of King Henry IV, against the multiplying gold and silver.'[14] Boyle thought he had got the knack ... but it didn't quite work out that way.

That of act repealing the ban mandated any alchemist who managed to make gold, to give it to the monarch! The next chapter will describe how this did happen just once in British history, nearly a century later.

Demise of the Seven-Metal Theory

"There are seven planets and also seven metals' wrote Paracelsus – although he knew of the existence of such semi-metals as antimony, arsenic, bismuth, and he even mentions zinc. Yet it would never have occurred to either him or anyone else of his time to put these on a par with the traditional seven.

Antimony for example was well-known in the sixteenth century; shiny, metallic-looking streaks of it were found in ores. But what they called 'antimony' was really the oxide of antimony. Metallic antimony was also well-known and was called 'regulus of antimony,' but it never occurred to anyone to regard it as more metallic in nature than what they called antimony. Not until the eighteenth century did this become evident: a dictionary of 1788 described how:

> *pure regulus of antimony is a bright semi-metal resembling tin or dusky silver.*

Ores of cobalt and nickel became fairly well recognised in the

---

[14] Hunter, M., 'Alchemy, Magic and Moralism in the Thought of Robert Boyle', Brit.Jnl. History of Science, 23 (1990), 404.

seventeenth century. They were named after the German Kobolds and Nickels, elemental beings of the mines! The classic German mining treatise *De Re Metallica* by 'Agricola' has been ridiculed on account of its description of the harassment which miners received from these sprites. Miners had to work under damp, dark and arduous conditions, at times with noxious vapours about. Nickel was first isolated in 1751, cobalt in 1733.

One popular textbook was the *Cours de Chemie* by the Frenchman Nicholas Lemery. Written in the late seventeenth century, it continued to be reprinted right through the first half of the eighteenth century, in twenty-three editions. It took the whole subject out of the smutty workshop of the alchemist and into the boudoir of high fashion. In it the number of metals remains unequivocally seven, but doubt has crept in as regards their planetary affinities. These appear as the old-fashioned beliefs of yesteryear, now being replaced by the new atomic philosophy. Of silver, for example, which he calls *Luna,* Lemery says

> It is called by the name of the Moon, as well from its Colour, as from the Influences which our Forefathers thought it received from the Moon.

Bismuth Lemery describes as 'an imperfect Tinn', a form of Marcasite, and also includes zinc in this category. Their properties are halfway to being metallic. Chemistry was reinterpreted so that all of the qualities and attributes of the preparations described were due to the shapes and sizes of small unseen particles of matter, atoms. Atomic theory gave the death-blow to the old doctrine, appearing as an alternative explanation for the properties of metals.

As the old theory went out of fashion, how was one to decide which substances were metals? An interesting judgement comes from a French chemist Pierre Macquer writing in the 1770s. The essential properties of metals he believed to be malleability and resistance to destruction by fire. The 'perfect ' metals had both of these attributes, and these were gold, silver and platinum. 'Imperfect' metals had only the first of these attributes: copper, iron, tin and lead. They would oxidize in a fire. Other 'metallic substances' did not meet either of these criteria: they were bismuth, zinc, regulus of antimony, regulus of cobalt and regulus of arsenic. (it is perhaps surprising that he should classify zinc and cobalt as merely 'metallic substances' along with the semi-metals.) Mercury he classed separate from the others, being still somewhat regarded as their parent.

Histories of science scarcely treat of the seven metal theory, except as some strange mediaeval superstition. Nobody would guess from them that it was a dominant chemical paradigm throughout most of the time that the metals have been known to humans.

Arrival of the New Metals

The notion that there existed just seven metals endured right through the scientific revolution of the seventeenth century, and up to the latter half of the eighteenth century. A new metal then appeared on the scene from the gold mines of Columbia, a 'white gold'. This was platinum, first reported in the year 1748. In 1750 notice of the new 'semi-metal' was given to Britain's Royal Society, though with its inert character and very high melting point there was very little that could be done with it. It soon became known as 'the eighth metal.'

Why was it the newcomer from Columbia which finally broke the sevenfold spell, and not zinc? Zinc had been known for two centuries, ever since it was first mentioned by Paracelsus, and its main ore (calamine, or zinc sulphide) had been used for making brass for two millennia.

This puzzle reminds us of the experiential criterion that once prevailed for the category, 'metal': it meant a substance that could be hammered into shape without shattering, that could be polished to a shine and heated without going up in smoke. Could an alchemist of old have heard the modern definition - viz, a metal as an element possessing free electrons able to float about in its atomic lattice - he would have regarded it as alarmingly unconnected with the world of experience.

Platinum was experienced as more metallic than zinc, even though it was totally unreactive. Zinc had been scraped off the roofs of lead refineries for centuries, just as bismuth had long been gathered in the course of mining silver; but zinc was far too volatile and combustible to be accepted as a metal, and semi-metals such as arsenic, antimony and bismuth were not felt to have properties comparable to the 'true' metals.

Mediaeval samples of brass often show thirty to forty percent of zinc in them, but despite this it was firmly believed that brass was merely copper 'tinctured' with calamine. Brassmaking simply involved heating copper with calamine. The ancients well understood that bronze was an alloy, because they came to Britain to get the tin to make it, but - so unshakeable was the sevenfold doctrine - no-one suspected that the same applied to brass. The notion seems first to have dawned upon J.R.Glauber in the late seventeenth century. As an alchemist Glauber believed totally in the traditional correspondences, but as an early industrial chemist he realised that purer brass could be prepared by first obtaining the 'immature' metal zinc and then fusing this with copper. His English counterpart, the chemist Robert Boyle, never suspected that brass was an alloy of two metals. The new method of brass manufacture from copper and zinc was being used here and there at the beginning of the eighteenth century, however it remained an industrial secret. Not until the latter half of that century did it become widely known.

Once electricity became available, other metals too chemically reactive to isolate by chemical methods began to appear. In 1807 Humphrey Davy produced potassium by electrolysis, this being the first of ten new elements he came up with. Uranium was discovered in 1789 by the German chemist Klaproth in the form of its oxide,[15] and decades later the metallic form was isolated.

The old names of chemical compounds continued on through the eighteenth century, though the basis for belief in correspondences was fading. Silver nitrate, the best-known salt of silver, was called 'Lunar caustic' and ferrous sulphate was 'Vitriol of Mars.' The 19th century saw the rise of a 'rationalised' nomenclature which finally eliminated the planetary names. For example, an 1826 chemical dictionary noted, under 'nitrate of Silver', "formerly called Lunar Nitre, Lunar Crystals or crystals of silver, and when fused Lunar Caustic."[16] In the nineteenth century, Dalton's new system of nomenclature cleared away the last vestiges of a belief, deriving from sky gods of ancient Babylon.

****

---

[15] Alison Davidson, *Metal Power* p.60. I tried in vain to find the date when Klaproth first isolated 'klaprothium', as he called his first uranium sample.
[16] Otley *Dictionary of Chemistry*, London 1826.

9. WHEN ALCHEMISTS MADE GOLD

Through five centuries of European history, fabulous moments are on record, when alchemists allegedly made gold. Gold was believed to be the most 'perfect' metal. Mother Nature worked underground – it was believed – causing ordinary metals in the Earth to 'mature,' so they would end up eventually as perfect of metals. This was an organic view, whereby Nature *aspired towards* perfection. In his crucible, the alchemist endeavored to imitate Mother Nature – just, speeded up a bit.

I found a collection of historic dates when the alchemists of old were supposed to have achieved this, ranging from the fourteenth to the eighteenth century. They started off with mercury, most often. The atom of mercury differs from that of gold, by only one electron! Yet gold is far heavier –

	Atomic No.	Density
Gold	79	19 g/cm^3
Mercury	80	13 g/cm^3
Lead	82	11 g/cm^3

Books usually aver that lead was the starting-point for the great, alchemic endeavour, but I did not find this. No wonder they preferred mercury to lead, there was only one electron difference! (that is the meaning of their atomic numbers 79 and 80 being next to each other) But, here we are *not bothered* with the supreme question of whether the alchemists 'really' made gold – that is too ultimate an issue, or it may be metaphysical even. Instead, we examine the *quality of time* when success in this fabled endeavour was recorded. Were any special celestial aspects present in the heavens, more often than one would expect by chance, at such moments? We aim to discern whether these events shared in common any particular quality of Time.

We have limited this inquiry to such occasions as were witnessed, that is to say, on which the alchemist was not alone: these are *witnessed and dated* alchemical goldmaking moments. By this approach we may hope to avoid the futile question of whether the alchemists 'really' made gold. The times when such events were recorded could well have in common some special qualities: there was for example a tradition for lead being the *prima materia* from which gold was created, in which case would one expect strong aspects to Saturn? Or, would Mercury play a key role on days when the 'Hermetic Art' was being consummated?

Altogether *nine* such moments were collected.[132] Several such moments have been recorded in the 20th century, which might be findable, however I did not quite manage that.

1. Flamel, 1837, Paris

In the 14th century, after 'years of unremitting labour', the French alchemist Nicholas Flamel recorded how he finally prepared the 'elixir':

> ...I made projection of the Red Stone upon half a pound of mercury, ... the five-and-twentieth day of April following, the same year [1382] about five o'clock in the evening; which I transmuted truly into about the same quantity of pure gold, most certainly better than ordinary gold, being more soft and more pliable...I had indeed enough when I had once done it, but I found exceeding great pleasure and delight in seeing and contemplating the admirable works of Nature. (Holmyard, p.245)

The chart of that time shows the the golden Sun conjunct Mercury within a 5° orb, and leaden Saturn (to 6°), themselves 1° apart. Pluto, the 'lord of transformation' conjoins Mercury to 8'. The red planet Mars casts a trine to this four-planet stellium.

---

[132] Sources: 'The Gold Makers' by K.Doberer(1948), 'The Secret Tradition in alchemy' by A.E.Waite (19), 'Alchemy' by E.J.Holmyard (1957), 'Alchemists and Gold' by Jacques Sadoul (1972), and 'The Arts of the Alchemists' by C.A.Burland (1967) The date for Edward Kelly comes from the diaries of John Dee, while those for Hevelius and the Guildford alchemist Dr James Price appear in their own publications. Where more than one such date was available, as for the alchemists Seton and Price, only the first was used. The original version of this article, published in Astrology (UK) Summer 1992, had only seven golden moments; it lacked that of 1716. (Jacques Sadoul was a pseudonym, and the English translation *'Alchemists and Gold'* was published by Neville Spearman in 1972. Hermetically enough, no copy of this opus exists in the London Library, nor it seems in any other London library.)

2. Kelley, 1586, Trebona in Bohemia

In the mid-19th century a private diary of Doctor John Dee came to light, 'written in a very small illegible script on the margins of old almanacks.' The diary recounts that learned doctor's journey to Bohemia, in the company of Edward Kelley. In the year 1588, the Elizabethan courtier Dyer received from Dee the news that his colleague 'had at last achieved the secret of the ages, that Kelley could indeed transmute base metals into gold.' This news brought Dyer to Prague later in the year to see for himself how matters stood.

Dee's diary for 1586 tersely records some stages of the Work: 'March 24th, Mr K. put the glass in dung.... Dec 13th, Mr E.K. gave me the water, erth and all.' Then, on 19th December 'novi kalendarii', meaning the Gregorian calendar, at Trebona, in the castle of Count Rosenburg,

> 'E.K. made projection with his powder in the proportion of one minim upon an ounce and a quarter of mercury and produced nearly an ounce and a quarter of best gold; which gold we afterwards distributed from the crucible.'[133]

On that day the Sun conjoined Mercury (2°), trined Saturn (1°) and squared Mars (4°). Dee was undergoing his second Saturn-return, (5°) synchronous with an exact fifth Jupiter return (1/2°). His Saturn therefore received the trine of transiting Sun (3°) and transiting Mercury (1°) on that day, while his North lunar Node was conjoined by them.

Dr Dee remained ignorant of how the process had been achieved, until the 10th of May 1588, when his diary states: 'E.K. did open the Great Secret to me, God be thanked.' On that day the Sun exactly conjoined Saturn, and Mars (1°), while the Moon was Full. It seems an appropriate day for Dee's insight, whatever that was.

To Dyer, Kelley later wrote, recalling

> what delight we took together, when from the Metall simply calcined into powder after the usuall manner, distilling the Liquor so prepared with the same, we converted appropriat bodies (as our Astronomie inferiour teacheth) into Mercury, their first matter.

---

[133] *The Private Diary of Dr John Dee*, p.22, Ed. J.A.Haliwell, Camden Society, 1842

Where do we find a modern chemist recollecting the delight he took in a chemical operation? ('The Private Diary of Mr John Dee', p.22-3)

3. Seton, 1602, near Amsterdam

Jacob Haussen witnessed the Scotsman Alexander Seton making gold from lead, at Enkhuizen near Amsterdam. Seton engraved upon it the date and time (N.S.): 13 March, 1602, at 4 pm. The Sun was conjunct Mercury and trine Saturn, both to 1°, square the nodes (4°), and semisquare Uranus (8'). Neptune held the ascendent with Saturn at the I.C. Venus was just setting, in opposition to rising Mars (1°). (Sadoul, p.119)

4. Richthausen, 1648, Prague

In the city of Prague, in 1648, the alchemist Richthausen performed a celebrated transmutation in the presence of Emperor Ferdinand III: ' with one grain of the powder provided by Richthausen, two and a half pounds of mercury were changed into gold. To commemorate the event the Emperor had a medal struck of the value of 300 ducats.. The inscription read (in Latin), 'The Divine Metamorphosis, exhibited at Prague, 15 January 1648, in the presence of his Imperial Majesty Ferdinand III.' On that day the Sun was trine Saturn (1°) and conjunct Mars (3°), and Mercury was conjunct the South Node (4°). The Sun conjoined the natal Saturn of Emperor Ferdinand who, at thirty-nine years of age, had reached his Uranus-opposition (22'). (Holmyard, p.129)

5. Helvetius, 1667, The Hague

Helvetius was Physician to the Prince of Orange. At the prompting of his wife, on the morning of 19 January, he melted lead and sprinkled over it some powder, as directed by the stranger who had given it to him. He recorded his wonder at seeing the gold: 'Yea, could I have enjoyed Argus's eyes, with a hundred more, I could not sufficiently gaze upon this so admirable and almost miraculous a work of nature.'[134] On that day, the Sun made a multiple conjunction with Mercury, Saturn and Neptune, all within 6°. The event generated widespread interest. Spinoza came to inspect the crucible and declared

---

[134] J.F.Hevelius, 'Golden Calf', reprinted by the Alchemical Press, 1987, US. Dr James Price, *'An Account of Some Experiments on Mercury...'* 1782.

himself convinced. (Holmyard, p.266)

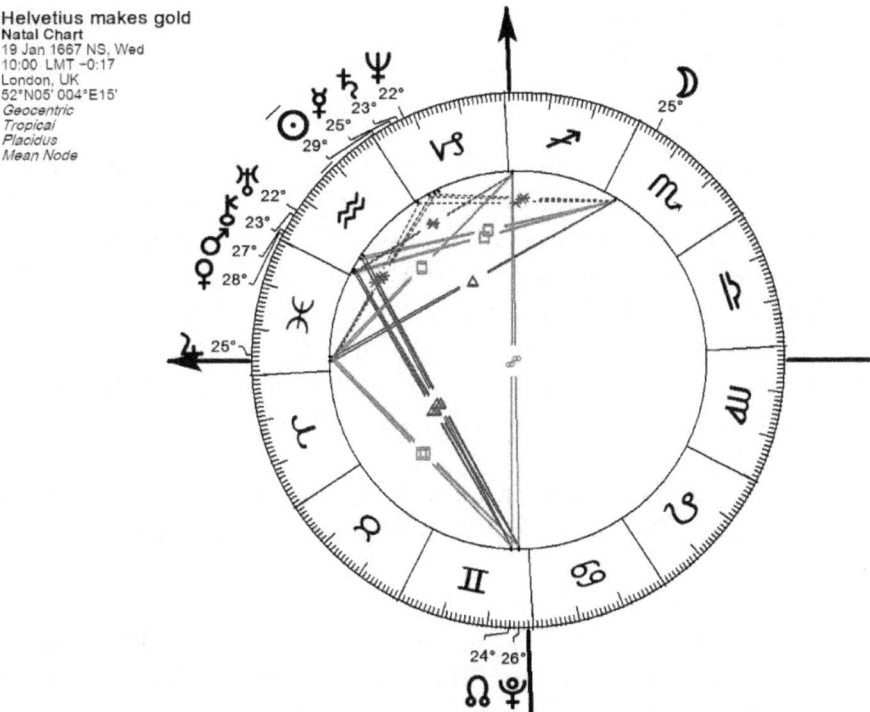

6. Böttger, 1701, Germany

'At the end of a good supper on October 1st, 1701, the apothecary Zorn, after some encouraging glances from his faithful wife, invited journeyman Böttger to give at last a demonstration of his skill.' Four persons believed that they saw, at the house of Herr Zorn, silver become gold. In the heavens the new Moon was but 4° from the Sun. (Doberer, p.234)

7. Lascaris, 1709, Germany

Little is known about the fabled figure of 'Lascaris', except that he was ascribed as the source of the 'projection powder' used by various alchemists, and that his name derived from a German noble family. On the 16th February 1709 in the evening, 'he is believed to have changed mercury into gold and gold into silver, a double transmutation.' The event was performed near Lissa and witnessed by Liebnackt, Counsellor of Wertherbourg. The story as told by Arthur

Waite came from a German opus of 1832.[135] A highly empowered Sun then stood in a remarkable grand trine, in which four other planets were also involved; but as well as this, a Uranus-Pluto conjunction was then occurring, in close opposition to the Sun-conjunct Mercury.

8. 1716 Rhineland

The British Museum's numismatics department has a coin replica, of an original that was kept in Vienna. The front side of the coin shows a picture of the mythic figure of Chronos, with his scythe and hourglass, being transformed into Sol. On its back is inscribed in Latin, 'Chemical Metamorphosis of Saturn into Sol, that is of lead into gold, witnessed by many on 31st December 1716, procured by His Serene Highness Charles Phillip, Count Palatinate of the Rhineland, Elector of Bavaria ... minted in perpetual memory, and donated to posterity'.

That day saw a grand trine of Sun, Pluto and Neptune, with Saturn conjunct the North Node in trine to Jupiter. (Burland 1967)

9. Price, 1782, Guildford

James Price, Fellow of the Royal Society, was a wealthy man of high social standing, with an established reputation as a chemist. In Guildford, Surrey, he carried out seven alchemical projection experiments during May 1782, of which six were successful, causing 'an immense sensation.' The sixth of these was conducted on Saturday, 25 May, with three lords present, including Lords Onslow and Palmerston. The latter put half a grain of 'a certain powder of deep red colour' on to some heated mercury, and after a while it seemed to have turned to gold. A sample was sent to an assay-master 'recommended by the Clerk of the gold-smith's company' who reported the gold to be pure. It was then sent to an experienced

---

[135] A.E.Waite, *The Secret Tradition in Alchemy*, p.324; quoting from C.C.Schmieder, Geschichte der Alchemie 1832, 1927, p.481. Louis Figuier in L'Alchemie at les Alchemistes 1860 p.328 also described this event, but cites the date of 16th February 1704, ie a decade earlier, in the village of Asch sur l'Eger; Waite says he found Figuier's accounts of Lascaris unreliable and told in a 'pseudo-historical manner' (Waite, p.321). As the events happened in Germany, we may reasonably prefer the German version, not least because Schmieder troubles to state that the transmutation happened in the evening: it seems unlikely that this detail would be supplied, if the date erred by a decade.

goldsmith of Oxford, who said it was 'superior to Gold of the English standard.' In Price's initial experiments only a small fraction of the mercury turned to gold, for example 1/8 on the 9 May attempt, but on this occasion most of it did so.

A remarkable pattern built up in the heavens over May of 1782. The climax of Price's experiments came on 25 May, as a close Jupiter/Saturn conjunction opposed a close Mars/Uranus conjunction (W " X 2°), an image of cosmic tension, while Sun-conjunct-Mercury completed a grand trine with Neptune and Pluto. When the experiments started on 6 May Mars was still 9° from Uranus and the Sun stood in no particular aspect. The main thing of note at the start was Uranus' close opposition to Jupiter/ Saturn. Following the demonstration before Lords Onslow and Palmerston, a last trial took place on Tuesday 28 May. The fourfold opposition was still present as was the solar grand trine, but in addition the Moon came into conjunction with Jupiter/Saturn, enhancing that already tense opposition. King George was presented with gold from that experiment, and 'was pleased to express his approbation.'[136]

That event represents a climax and termination of the British alchemical tradition, so let us note that the Mercury of this chart was conjunct the Sun of the King within one degree as he received the alchemic gold (while that mighty opposition in the sky fell on the King's Neptune). Oxford University had awarded Price an MD 'on account of his labours,' although in the heat of the controversy later claimed it was for his earlier chemical work, not for his alchemy. The sober Royal Society members and his fellow-chemists remained skeptical, 'knowing the impossibility of his assertion ', to quote from their impartial report. Steps were taken to expel 'our Paracelsus of Guildford ', as he was called, from that august fraternity. 'Was ever a country more completely disgraced than ours has been by the conduct of our University?' complained one irate member to the Royal Society's President. Clearly, something had to be done.

---

[136] *Notes & Records of the Royal Society*, Vol.9 p.109-14 'The Last of the Alchemists' by H.C.Cameron. Also E.J.Holmyard, p.267.

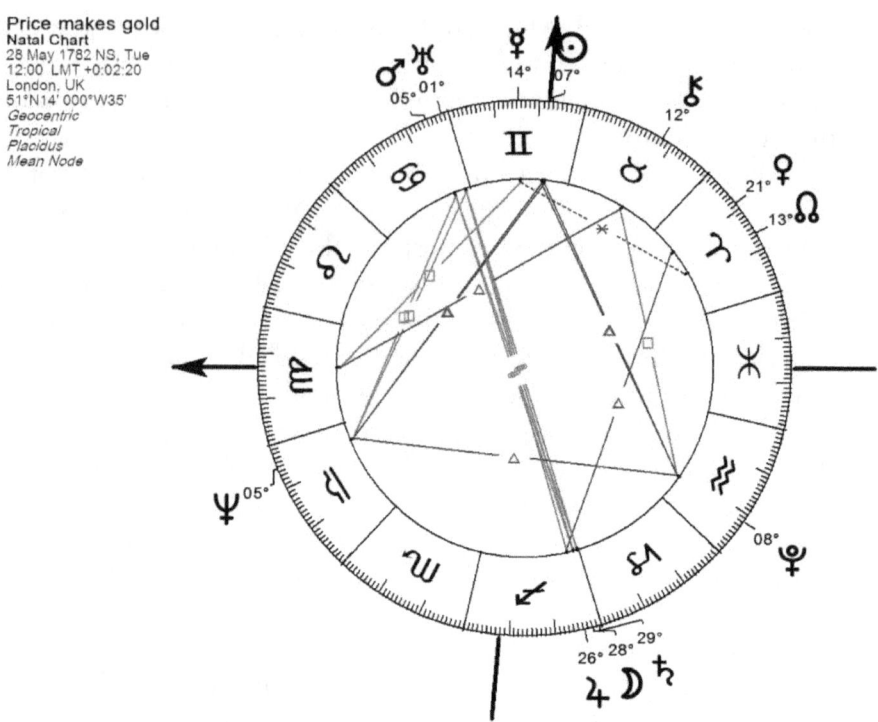

Requested to repeat his gold-making demonstrations, Dr Price excused himself on the grounds that his mysterious powder had run out: to make more, he explained, would be 'difficult, dangerous and injurious to the health', a phrase which doubtless left no-one much the wiser. He claimed he had sworn the proverbial vow of secrecy over the recipe, to an unnamed person who had given it to him. Eventually, the Doctor was prevailed upon to attempt a repeat of his earlier experiments, and a trio of Fellows duly arrived at his residence in Guildford in August 1782.

Price excused himself from their company for a moment that afternoon and drank hemlock in his kitchen. Returning, he died in their presence. The coroner passed a verdict of insanity. Neither the date of Price's birth or even of his death are now known, though he had been awarded an Oxford doctorate and elected a Fellow of the Royal Society. Don't even ask whether anyone preserved the alchemic gold which he gave to the king. When I lived in Guildford I sought for but could not find any record of where he had lived. He has been well and truly erased from history – 'The last of the alchemists' being his glorious epitaph.

Those who turned up on his doorstep were agreed on their axiom, whereby a phenomenon was only real if repeatable. Price knew otherwise, but could not explain this - at least not in their language, for whatever it was that he had done. Crushed by the new science, he chose death before dishonour. He took with him the untellable fact, the inexpressible story.

King George was going mad, due to medicine from his doctors: they were giving him antimony, which had traces of arsenic. Posthumously he was found to have three hundred times the acceptable level of arsenic in his hair. The going-mad king expressed his approbation for this priceless gift from Price, while the new, 'objective' Royal Society drove him to suicide.

So there is a bit of solar glory in this golden tale, even though it does involve hemlock ... and arsenic.

Planetary frequencies

The goldmaking dates

Date	Time	Alchemist	Metal	Conj	Place
25.4.1382	5pm	N. Flamel	Mercury	☉☌☿	Paris
19.12.1586		Kelly	Mercury	☉☌☿	Trebona
13.3.1602	4pm	Seton	Lead	☉☌☿	Amsterdam
15.1.1648		Richthausen	Mercury		Prague
19.1.1667	10am	Helvetius	Lead	☉☌☿	The Hague
1.10.1701	8pm	Böttger	Silver		Germany
16.2.1709	Evg.	Lascaris	Mercury	☉☌☿	Lissa
31.12.1716		?	Lead		Dusseldorf
25.5.1782		Price	Mercury	☉☌☿	Guildford

Collecting together all nine of these gold-making dates, the Table shows that they shared no less than *six* Sun-Mercury conjunctions, chiming to within five degrees. Those events where mercury was the *prima material* have been highlighted. Taking these nine dates, I added up the planetary conjunctions, oppositions and trines present in them to 5° orb, and scored them per planet, to give the following:

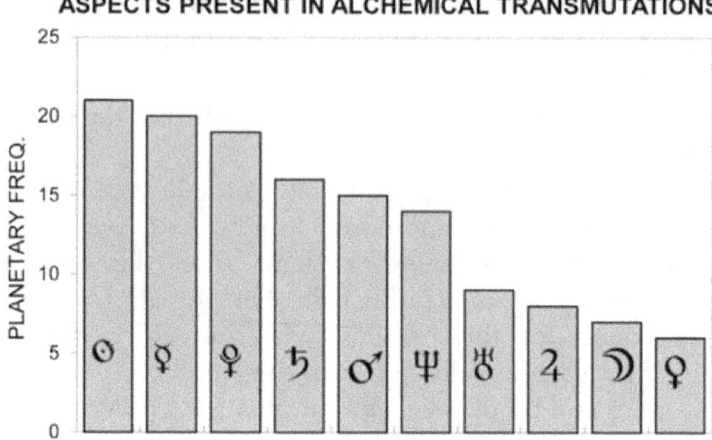

The golden Sun came out top, making altogether 21 of these three major aspects in the 'goldmaking' charts. Right at the bottom of the chart is Venus, having made only six aspects. The Moon is almost bottom of the list, and so there is by any standards a dramatic and quite alchemical polarity between these two luminaries: the pale, watery Moon was definitely of no use during these fiery moments! She had only one-third as many aspects as did the brilliant Sol, who fully comes up to expectations in this chart! The two feminine planets Venus and Luna skulk at the bottom of the planetary-frequency list, they are no help here.

For a given planet, by chance one would expect nine such major aspects within the group;[137,9] ie one per chart. Thus the last three planets, Jupiter, Venus and the Moon, have scored somewhat below chance level. Overall, there was a huge 50% excess of these aspects in

---

[137] A 5° orb gives 10° of the ecliptic per aspect, and as there are two possible trine positions and only one for conjunctions and oppositions, where each planet can form aspects to nine others, each chart will have an expected aspect frequency of $10(1+1+2) \times 9/360 = 1$ per planet. For the Sun, due to the aspects it cannot form with Venus and Mercury, the equivalent figure is $\{10(1+1+2) \times 9-6\}/360 = 0.95$ expected aspects per chart. Taking just the three aspects between SA, SU and ME for all nine charts, their expected frequency would be $10(1+1+2)3 \times 9/360 = 3$.

9) I checked all these figures using the Jigsaw astrology-research program. For more details on computing these expected aspect frequencies, see N.K., 'Investigating Aspects' in *Astrological Research Methods*, Ed. Mark Pottenger 1995 ISAR CA, pp.287-302.

the gold-making charts. This is highly unlikely to have arisen by chance. It is therefore evident that, overall, these charts had very strong major aspects.

Pluto is associated by astrologers with atomic transmutation: experiments in this realm began in the 1930s immediately following Pluto's discovery. It is most surprising to find this then-unknown planet second in the list of aspect frequencies after Sol. Its high score here provide an unexpected bridge between the alchemy of a bygone age and the atomic transmutations of the twentieth century.[138]

Mercury and Saturn both scored highly, at twice their chance level. On days when the Hermetic Art was being consummated, it seems appropriate that Mercury should score so highly. This was the most often-used metal for the transmutations, on five out of nine occasions, followed by lead. The aspects between Sun, Mercury and Saturn were:

1) ☉ ☌ ☿ 4°,   ☿ ☌ ♄ 1°
2) ☉ ☌ ☿ 2°, ☉ tri ♄ 1°, ☿ tri ♄ 3°
3) ☉ ☌ ☿ 1°, ☉ tri ♄ 45', ☿ tri ♄ 0°
4)   ☉ tri ♄ 49'   ☿ R
5) ☉ ☌ ☿ 4°,   ☿ ☌ ♄ 1° ☿ R
6) -   ☉-☿ 8°
7) ☉ ☌ ☿ 3°, ☉ tri ♄ 2° ☿ tri ♄ 5°
8) -   ☉-☿ 10°
9) ☉ ☌ ☿ 4°

The Sun cannot form trines or oppositions to Mercury or Venus, so has a slightly lower expected score. In other words, its excess is even larger than appears on this chart! Major solar aspects are present in this group at at least 150% more than would be expected by chance.

By chance one would expect about three such aspects to be present in such a group, but here there are fifteen! Mercury spends one-eighth of the time within five degrees of the Sun, so the likelihood of finding this conjunction in two-thirds of the charts is remote.

---

[138] See plutonium chapter, also Robert Chandler, 'Uranium, Plutonium and Black Alchemy' *Bulletin of The Company of Astrologers*, Summer 1995, 21-28.

Mercury's average angle with the Sun in this group was 5.3°, whereas normally it averages 15° I found, and that is a huge difference (sampling Mercury 'elongations' ie the ecliptic angle between Sun and Mercury, over the two yeas 2011 and 2012). The conjunctions are definitely being chosen in this group.

Claims that gold had been made, on these famous occasions, stirred up frenzies of public excitement. Strong solar aspects were found to be present at these moments.

Two of the goldmaking moments had Mercury moving retrograde, denoted by an 'R' in the Table. The two kinds of solar conjunctions made by Mercury, 'inferior' and 'superior', happen with equal frequency, the former when Mercury passes in front of the Sun and the latter when it is behind. The former is Mercury's closest approach, the only times when it goes retrograde (ie, appears to move backwards against the stars). The Table shows six conjunctions of Sun and Mercury, five superior and one inferior, with mercury not being the *prima materia* in the case of that inferior conjunction. This group comprises all dated, witnessed goldmaking events I found.[139] I am grateful to Scotland's alchemist Adam McLean for help in locating them.

---

[139] A source of further witnessed, dated, goldmaking moments could be *'L'Alchemie'* by Jacques van Lennep (Belgium), 1982. There may have been some 20th century goldmaking moments that I could not find.

10. THE KOLISKO EXPERIMENTS

I returned to port so battered by the tempestuous seas that I did not dare to await the opposition of Saturn with Mars, which generally brings storms and bad weather.

From the log of Christopher Columbus, Christmas 1502, during his fourth and final voyage to the Indies

Can metal experiments show the course of a celestial event? This question gripped me in my youth. In two books I gave accounts of the 'Kolisko' experiments,[140] whereby the reactions of metallic salt solutions are may vary with celestial events, for corresponding planets. These were done some decades ago, with little by way of replications since. I will here give an outline of what has been done since the 1920s, using quotes from others on the subject.

The Meetings of Mars and Saturn

<u>Published experiments done over Mars-Saturn conjunctions, 1952-2002</u>

Kolisko.	Dec 26, 1927	♂☌♄	Saturn und Blei	1952
"	Jan 17, 1934	♂☌♄	"	
"	Jan 27, 1936	♂☌♄	"	
"	Nov 30, 1949	♂☌♄	"	
Schwenk	Nov 30, 1949	♂☌♄	Secrets of Metals	1959
Karl Voss	Feb 15, 1964	♂☌♄	Neue Aspekte	1964
Drummond	Jun 16, 1970	♂☌♄	Astrological Jnl.	1976
Desbolles	May 4, 2002	♂☌♄	Sciencegroup.org	2002

At least seven conjunctions of Mars and Saturn, through eight

---

[140] NK, *Astrochemistry* 1984 and *The Metal-Planet Relationship,* 1993.

decades, have been recorded by various people in terms of changes in chromatography-type pictures (called, 'steigbild' in German) formed by precipitation of lead, iron and silver out of solution. There were two experiments done over the 1949 event: Kolisko's weighty *Saturn und Blei,* published in Stroud 1952, described one such experiment, while that by Theodore Schwenk in the Weleda laboratories in Switzerland has only appeared (that I know of) in Wilhelm Pelican's book, that was translated into English in 1973. I suspect I'm the first to notice this replication. The above table indicates the astonishing sequence of replications of this phenomenon.

A PDF of my hopefully-groundbreaking 1977 article *Chemical Effects of a Mars-Saturn conjunction* is available online http://www.astrology-research.net/researchlibrary/NKMarsSaturn.pdf , concerning the 1976 experiment I did with Michael Drummond. Lead and silver were measured spectrophotometrically as present in the filterpapers. One could see the chiming of that celestial event, through the changing lead levels they contained. Later on it dawned on me, that my doing this coincided with my Saturn-return at the age of 30.

For a summary of the various different Kolisko publications on this topic, see John T Burns *Cosmic Influences on Humans, Animals and Plants,* 1997, online. This reviews various articles and letters on the subject by Kolisko, myself and Agnes Fyfe (a Scottish lady who worked in Switzerland, in the Weleda lab).[141]

These Kolisko experiments have two credulity-straining features. The first is the claim made by Rudolf Steiner, to Kolisko when she was still in Germany, that: 'So long as substances are in a solid state they are subject to the forces of the earth, but as soon as they enter the liquid state, planetary forces come into play.' That is the essential theory involved in these experiments. Secondly, there is the strange fact that simple ionic reactions, in this case between silver nitrate and ferrous sulphate in solutions, are slow. Several minutes go by after mixing, before anything starts to happens. It's to do with colloid formation.

As Kolisko explained it, the principle involved 'etheric forces,' which somehow work through or express themselves in form. Thus,

---

[141] Burns' book is only partially online, as a Google book. A comprehensive online text would be desirable. Each of its paragraphs summarises a published report or book, outlining a new science of cosmo-biology.

the modern alchemist Adam McLean expressed the view:

> I believe that the work of Lilly Kolisko has provided for us a foundation stone upon which future alchemical experimentation can be built ... I hope that her work can be continued and extended as I can see that upon its foundations a qualitative science of the etheric forces can be built, a new alchemy and Etheric Science.[142]

That's optimism! Kolisko's first experience of that came in 1926, over a Saturn-Sun conjunction, when she was using her silver-iron-lead procedure she had just devised. She saw the change in pattern:

> An invisible hand had blotted out the working of the lead in my solution. The Sun had stood before the planet Saturn and here below on Earth the lead could not manifest its activity. When the Stars speak man must stand still in silent awe.

One regrets that she kept describing her work in terms of the stars, naming her publications as 'the working of the stars in earthly substance' - her work had nothing to do with the stars. This may have irritated people, as too did her calling herself 'Lilly Kolisko' when her name was Elizabeth.

Here is how she experienced the difference between the working of gold and silver solutions.

> In the case of pictures of silver we find that the wealth of forms is so great that it is impossible to present one picture only of silver ... In the case of gold, the colours are so rich that many pictures must be observed before we can realise the nature and character of the metal. The colours that make their appearance vary between pure yellow and dark violet.[143]

But, the basic experience of these Mars-Saturn experiments is in a sense depressing, which (I sometimes wonder) could be why one does not hear much about them: a heavy blackness appears over the filterpaper, blotting out the normal forms. To quote Wilhelm Pelikan in 1959 summarizing Theodor Schwenk's 1949 Mars-Saturn experiment:

---

[142] Adam McLean, http://www.levity.com/alchemy/kolisko.html 'Capillary Dynamolysis', reprinted from *The Hermetic Journal* 1980.
[143] Kolisko, *Working of the Stars in Earthly Substance*, 1928, Foreword.

But on the day of the conjunction, November 30th, at 4.pm, a completely unusual picture suddenly appears. The broad, heavy "lead formations" become pointed and narrow and are reduced in number; a strong blackening of the background appears. Until the following day the reactions continue in this abnormal fashion, then slowly return to their normal condition.'[144]

The chemist Geoffrey Dean, astrology arch-sceptic, who has dedicated his life to debunking the subject, took an interest in these experiments. He alluded to reproductions as having been performed by "various workers including Schwenk, Voss, and Landscheidt in Germany, Fyfe in Switzerland, Faussurier in France" and myself, and added:

> Most of these workers have tested only Moon/Mars or Mars/Saturn events using Kolisko's or Fyfe's methods but all have observed a general decrease in the number of forms at the time of the event. The results to date suggest that the decrease is asymmetric: the effect begins at or just before ecliptic exactness and is prolonged after exactness and thus seems analogous to the persistence of ripples when a liquid is disturbed.[145]

Let's quote from the report by Michael Drummond, after he had completed his astonishing sequence of Mars-Saturn experiments:

> Kolisko has shown the way toward a new chemistry, a chemistry full of life, because it is in reality the forces of life which reach right down into the mineral world, creating there a realm of inner music in tune with the harmony of the spheres above.[146]

Michael started his daily experiments on May 6th, with the conjunction chiming (this is called the moment of 'exactitude' on May 12th). For days the forms vanished, and then: 'On May 20th, eight days after the conjunction, a change occurred. Dark but strong forms emerged, even though it took over an hour for the first signs to appear.

---

[144] Wilhelm Pelikan, *The Secrets of Metals*, 1973, p.25
[145] Geoffrey Dean, *Recent Advances in Natal Astrology A critical Review* 1977, 233.
[146] Mike Drummond 'Cosmological influences in Chemistry', *Mercury Star Journal* Autumn 1977.

One could feel that things were returning to normal; that the air had changed.' After we'd measured the lead in his papers, he reflected: 'It is interesting to note that the amount of lead in the papers corresponds well with the observed darkness, mentioned earlier. Both reach a peak seven days after the conjunction. So, although some effect of the conjunction can be seen two days before, when Mars and Saturn are still one day apart, the climax occurs when they have passed and moved three degrees apart (by geocentric longitude).'

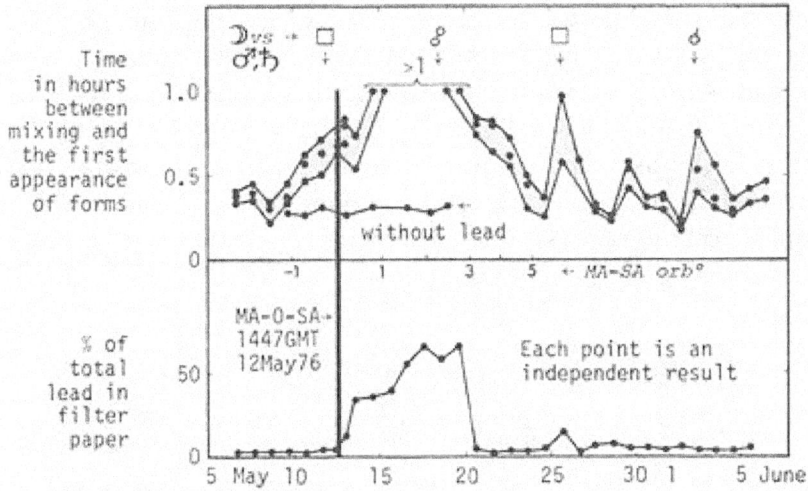

Figure: graphs of Drummond's 1975 Mars-Saturn conjunction experiment, from Dean 1977

During the Mars-Saturn conjunction, for one day before and six days after, darkness-without-form appeared on all of his filterpapers. I remember it, it was so amazing. He gave the filterpapers to me, and I cut off one centimetre strips from their edges and measured the silver and lead which they contained. Inspired by this, he did similar time-experiments over three other Mars-Saturn celestial events, as shown in his table. It's hard to keep the filterpapers as the silver goes brown with age.

Geoffrey Dean's graph summarised my measurements on this epic time-series, which Drummond performed in his 'home lab' in Henley. The thick line shows the chiming of the big conjunction of 12th May, with Moon-phase events included by way of showing how silver levels in the papers responded to them: Sun-moon quarters, Full

Moon and New. Three filterpapers were done each day and in the top graph you can see the time for first precipitation to appear in each paper as noted by Drummond, while the lower graph shows my measurement of lead in the paper: lead sulphate should be totally insoluble, as shown before the conjunction, it's not supposed to rise up the filterpaper. Compare these graphs with the days of disturbed form as noted by Mike Drummond.

He is the only person to have done these experiment with trine, square and sextile aspects (between Mars and Saturn), here is how he described what he found: 'The colour of the deposits on the papers remains normal, a light brown, during the sextile and trine aspect but instead of the usual rounded forms there are just a series of black dots, or seeds, from which these would normally grow. One could describe these forms as being stunted in their growth. The square aspect is stronger in its effects. As at the conjunction, a formless grey film appears on the papers, although of a lighter shade.'

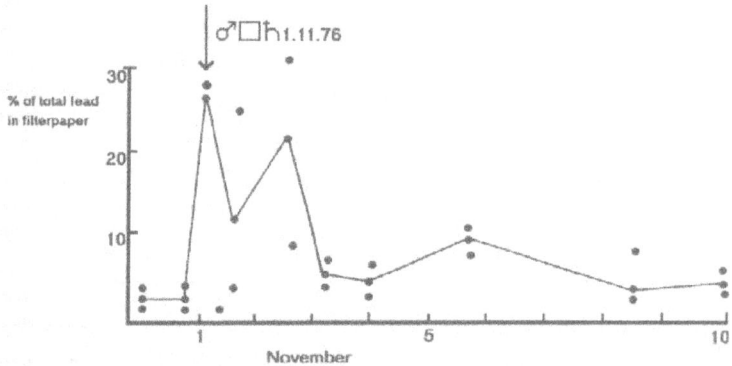

Figure: Lead in filterpaper for Mars-Saturn Square: Drummond 1976

The graph shows how the lead in the filterpaper was around 20% of the total, during that square aspect. I hasn't risen up quite as much as for the conjunction. Each point on that graph represents a measurement, one per filterpaper. This time it wasn't me that did the measurements: I alas misused the atomic absorption spectrophotometer in the City University but Dr Benbow could still allow me to use its laboratory. So I chopped off one centimeter strips from Mike Drummond's Mars-Square Saturn filterpapers and treated them with acid to dissolve the lead, then gave the batch of test-tubes to my colleague R.M. who was working at the Government Chemist laboratory, and he did the measurements.

Mike Drummond summarized what he had experienced: 'We have seen how the formative forces of Saturn, working in lead, are revealed through silver. A deterioration of these forces is evident when Saturn is either in line or harmonically related to other planets, of which examples have been given when Mars is the other planet involved.'

The table shows 'days of disturbed form' which he published summarising his experiments, over Mars-Saturn aspects.

Days of disturbed form in the Drummond Mars-Saturn experiments, 1976

Day		-1	0	+1	+2	+3	+4	+5	+6
Conjunction	12 May	x	x	x	x	x	x	x	x
Square	1 Nov	x	x	x	x	x			
Trine	13 Dec	x	x	x	x				
Sextile	11 Sep		x		x	x			

Let us heed Mike Drummond's conclusion: 'In the real world all is change. To maintain, as most chemists would, that the chemical nature of a substance always remains the same under unchanged physical conditions is not only untrue, but creates a powerful image of a dead, mechanical world.'[147]

---

[147] Compare Simon E. Schnoll, *Cosmophysical Factors in Stochastic Processes* 2012 (online) translated from the Russian of 2009. One commentator noted: "Since 1979 Simon Schnoll had investigated radioactive decay processes more closely, at laboratories thousands of kilometers apart, using automatic measuring devices. These tests showed that the distribution histograms of the decay rates of different radioactive samples had identical fine structures. The similar histograms generally repeated every 23 hours 56 minutes, every 27.3 days and every 365 days so that Schnoll concluded that the fine structure of the distribution of measuring results for processes of most diverse kinds is probably determined by some cosmological factors." These periodicities in radioactive decay patterns thus define the fundamental *sidereal* cycles which we experience: the rotation of Earth against the stars in 23 hours, 56 minutes, the Moon's rotation in 27.3 days against the stars, and Earth's sidereal rotation around the Sun in 365.2 days. I published a study "The Diurnal Cycle and Chemical Reaction Rate" at a symposium held in 1992, comparing these studies with what Kolisko had found. As professor John T. Burns kindly commented: "Finally, in one of the few references to the chemical tests of Kolisko in the mainstream literature, Kollerstrom mentions that Kolisko found

We turn next to an image formed using only silver and iron salts, no lead – a much quicker reaction. It shows some rather heavier than usual forms created as solutions of iron and silver react together while rising up a filterpaper. This one was made immediately before a conjunction. I did this 1975 experiment with Charles Harvey and Frank Hyde (both dead now) in a school outside London. I had insisted it had to be away from traffic. It was filmed by a BBC 'Horizon' team. My writeup commented:

> The Horizon team for their astrology program filmed (in 1975) the Moon-Mars conjunction, which in fact went rather well, with most impressive forms tuning up for the occasion, and being replaced by a marvellously delicate pattern, identical in all three filterpapers, shortly after the conjunction. The BBC decided not to show it however and destroyed the film.

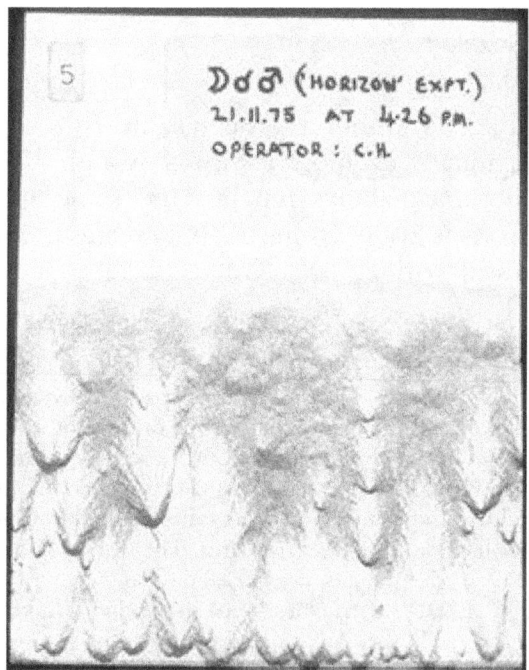

Figure: *Steigbild* using 1% solutions of iron sulphate and siver nitrate, just before a Moon-Mars conjunction

differences between day and night in silver precipitation and in crystallization tests." (Cosmic Influences on Humans, Animals and Plabts" 1997 Online.

Alas! I was told that the BBC team were 'stunned' by what they saw. *Therefore* they could not show it – the BBC as usual works by pretending to be interested and sympathetic to get the film, then doing a hatchet-job. So, for the record, I testify that me, Frank Hyde (he was a Fellow of the Royal Astronomical Society, and had invented something about radio telescopes), Charles Harvey and the BBC team were impressed that the Kolisko effect had actually worked under exact test conditions.

In consequence of this BBC interest, a chemist Dr Benbow from the City University in London had agreed to collaborate. He and Frank Hyde participated in witnessing another Moon-Mars experiment at Frank's flat in Muswell Hill, done in the small hours of the night: I insisted that his would be necessary. I believe everyone was convinced that it had 'worked.'

Thereby I gained access to an atomic absorption spectrophotometer, at the City University lab and was able to measure the silver and lead on the filterpapers, which is how the graph from Dean above came about. In my old notebook this analysis is dated 30.7.76 to 12.8.76: it began with a Sun-Saturn conjunction, on my Saturn-return, ending within arcminutes of my natal Saturn position. At that time I had just taken a science degree and had no particular belief in astrology, so I was rather confused by this.

Various people did these Kolisko experiments and obtained positive results which were published: Zach Matthews (Moon opposition Saturn, August 1976), Richard Moxon (Moon conjunct Saturn, April 1976), Frank Hyde (with me and Dr Benbow, Moon conjunct Mars, January 1976) me NK (e.g., Moon conjunct Saturn and Moon opposition Saturn, June 1970), Mike Drummond, and Charles Harvey at the 1976 AA Conference (Moon conjunct Mars August1975). You could say that nothing much happened as a result of this great endeavor, it all faded into oblivion: but, yes, we 'proved astrology' if anyone ever can.

The Mars-Saturn conjunctions have not usually exerted so long-lasting an effect. The above-quoted experiment by Theodore Schenk of 1948 had the darkness-without-form lasting a couple of days; as likewise in Guy Desbiolles' Mars-Saturn experiment in 2002 – fifty-four years later – the blotting out of form lasted two to three days: http://www.sciencegroup.org.uk/kolisko/desbiolles.htm .

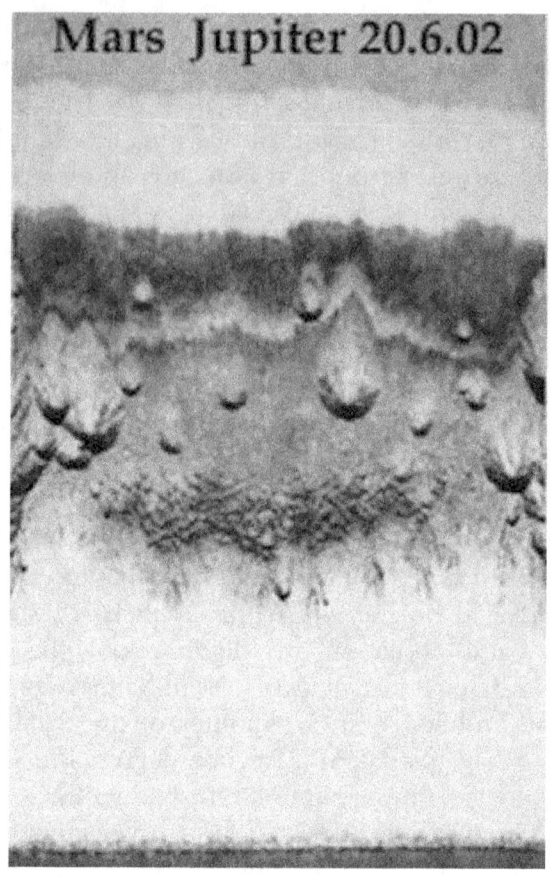

Steigbild using solutions of iron, silver and lead, ten days before the conjunction.

For his Mars-Jupiter experiment however (3rd July 2002) Guy found: "The fading of the structures is ... spread over a longer period i.e seven days centered on the day of the conjunction." An image of his a couple of weeks before the event is here shown. His are the only 21st-century Kolisko experiments, at least reported. He is a chemistry teacher in Switzerland. Above is a photograph taken by him, from the start of a Mars-Jupiter conjunction sequence he did in June-July 2002. The line at the bottom of the filterpaper shows where it stood in the petri-dish, where the solutions are mixed: the pattern results from the slow precipitation. This becomes far slower if the lead salt is added.[148] He wrote to me back in 2005 saying: "The students implicated in the experiments with Moon-Mars conjunctions have been able to reproduce the usual effects though not for each conjunction.' They do repeat, but not every time.

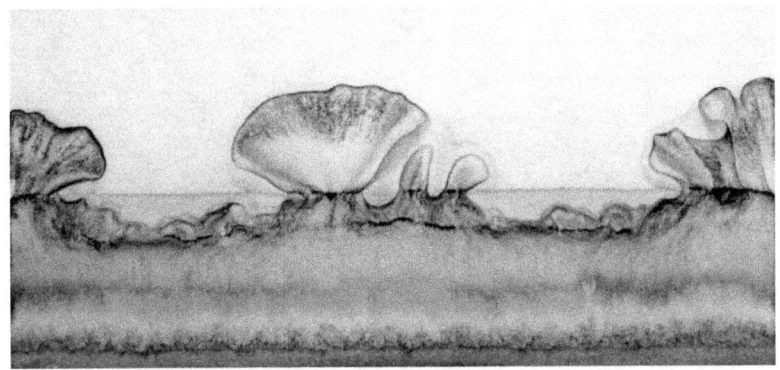

Figure: poppy juice with silver solution, over an eclipse.

Guy has also done a gold colloid experiment over a 2005 solar eclipse. He seems the only person to have continued Kolisko's gold experiments, while my endeavors here made little progress.

Here are a couple of plant-sap silver images - 'steigbild' is the German word - kindly provided by Guy. The first is poppy, and he did it over the day of a solar eclipse. The horizontal line shows how far up the poppy juice-extract rose in the filterpaper, then later on silver nitrate was risen up through it.

The second image was done differently, with the filterpaper flat and the juice extract fed in through the centre. Carrot juice was used, and the circle at the centre sows how far it spread out. Then the next day silver spread out through it, making these splendid forms! The general idea here is that this is an organic or 'bio-dynamic' carrot, and that its image shows 'healthy' life-forces or whatever you want to call it, whereas a chemically-grown carrot would give a duller form. Let us hope that there can exist a quality-testing centre using this method in conjunction with others, to ascertain that vital but hard-to-define concept, of food quality.

---

[148] For Kolisko's Tin-Jupiter experiments (*Der Jupiter und das Zinn: Sternwirken in Erdenstoffen*) see review in Burns 1997.

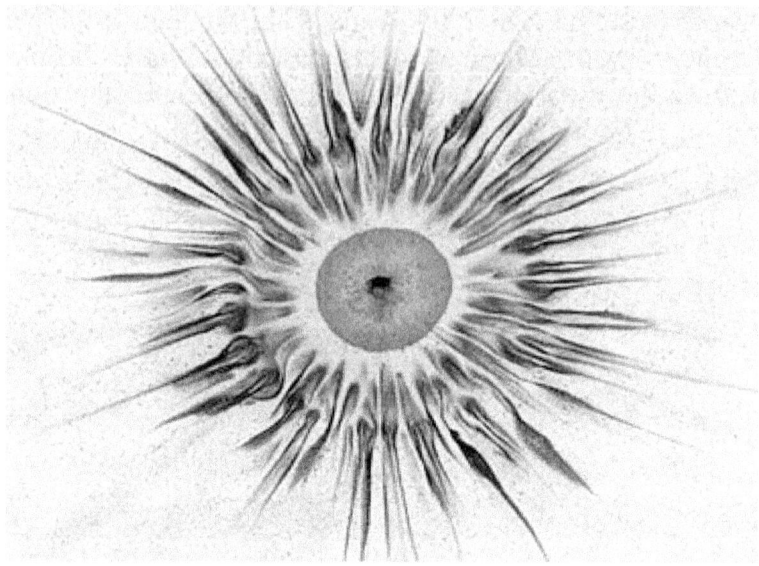

Figure: carrot juice followed by silver solution, on a flat filterpaper

Chemistry as something which we get to experience has greatly vanished from our society, now being controlled by huge corporations. I'd like to conclude with words from an autobiography of Frank McGillion, describing an emotion which readers are most unlikely ever to experience: delight at entering a chemistry lab. As if from a vanished era, we read -

> He pulled open the door and stood aside to let me in.
> Heaven! No, even better! It was beyond my wildest dreams.
> There were test tubes and retorts, conical and volumetric flasks, measuring cylinders, pipettes, beakers and droppers that tapered to perfectly flattened ends. There were real-sized Bunsen burners, jars of chemicals, balances and test-tube racks; there were big brown bottles, rubber tubing, a fume cupboard, wall charts of the Chemical Elements and dark rubber bungs with glass tubing sitting neatly in holes in their tops. It was perfect – just perfect.
> 'So what do you think?' asked Big Bill, following me in.
> 'It's very nice, sir,' I said, calmly, understating my feelings as ever.[149]

---

[149] Frank McGillion, *On the Edge of a Lifetime*, 2002, p.28

Websites:
www.skyscript.co.uk/metal7.html,
www.sciencegroup.org.uk/kolisko/
www.levity.com/alchemy/kolisko.html
www.spirit-web.org/mysteries/astrology/the-correspondence-of-metals-and-planets

11. PLUTONIUM, PLUTO'S ELEMENT

Underground millionaire Pluto lord of Death
Allen Ginsberg, Plutonian Ode[150]

From 1940 onwards the unnatural new element plutonium was made and bred under a cloak of absolute secrecy. For two years, *it had no name.* Ten years after the new planet Pluto had appeared in the heavens, this unnatural element appeared under the worst possible circumstances, as the world lurched into total war. This was part of the wartime 'Manhatten Project.' As Pluto in mythology wore a helmet of invisibility, so plutonium is an element which none of us ever see. The world first heard about it when it exploded in New Mexico, turning the desert sand to glass.

Pluto's turning-point

The new planet Pluto appeared in 1930 conjunct its own node. Its orbit makes a steep angle to the ecliptic so this was quite a pronounced event; therefore its appearance was powerful, as if some Hades-type principle had emerged into the light of day. Fascism grew in Europe, jazz and psychoanalysis became popular, a new era of invisible astronomy began using infra-red telescopes, antimatter was detected (1932), massive organized-crime gripped American cities, and the first zombie and vampire movies appeared.[151]

In 1932 artificial transmutation began, and then in 1942 atomic energy was unleashed. The victors of World War II assumed that it was OK to target missiles on the cities of other nations, and for the

---

[150] Allen Ginsberg, *Plutonian Ode*, 1982 City Lights Books San Francisco, gives date of completion of poem as 14th July1978.
[151] The films *Dracula,* 1931 and *Vampyr,* 1932, the first zombie film *'White Zombie',* 1932, also, the horror classic, *'Frankenstein'* in 1931.

dreadful new weapons, Pluto's element became the trigger. Tension heightened until, in the 1980s, survey showed that a *majority* of Britons were expecting thermonuclear conflagration.

The necro-technocrats were in control. Hidden missiles deep in their silos, which we never saw, threatened to emerge into the light of day, and 'cruise missiles' were wheeled around Europe - plutonium-tipped, they were supped to 'melt into the countryside.'

The '80s were stressful because Pluto had entered within the orbit of Neptune. It came nearest in 1989. Its orbit is strongly elliptical, so this nearest approach (its 'perihelion') was quite marked. Once that distant sphere started to recede, the tension abruptly vanished: we could all forget about the terror of annihilation, and the shadow of the Bomb faded into yesterday's memory. The plutonium-crazed history of the nuclear arms race thus appears as framed by two events: Pluto crossing its node in 1930, then reaching its perihelion in 1989.

A new god of darkness

The Bard surveys plutonian history from midnight,
Lit with Mercury Vapour streetlamps till in dawn's early light
He contemplates a tranquil politic spaced out between Nations'
thought-forms
Proliferating bureaucratic & horrific arm'd,
Satanic industries projected sudden with Five Hundred Billion Dollar Strength ...[152]

Allen Ginsberg, Plutonian Ode

When Uranium was discovered in 1789, it was named after the new planet Uranus, found in 1781. No further elements beyond Uranium were discovered until after Pluto's appearance. Uranium was the ninety-second element in the Periodic Table of elements, and when the next two elements in this sequence turned up, elements 93 and 94, there was a kind of inevitability about their naming. They had to be named after the next two planets, and so were called neptunium and plutonium - both found at Berkeley using the new cyclotron. Glen

---

[152] But N.B., In Canada, nuclear reactors have successfully remained unlinked to any military program.

Seaborg who named the new element had (it hardly needs saying) no inkling of the awful symbolic appropriateness of the name he was giving. He would have guessed it would be fissile, that's about all.

Within the core of nuclear reactors, a transmutation-process goes through the sequence of the outer planet-names:

Uranium 238 to Uranium 239 to Neptunium to Plutonium 239

I tend to visualise the uranium cycle in terms of the four elements:

> *Earth*: Uranium is mined;
>
> *Air*: a winnowing separates the isotopes, the fissile U-235 from the denser U-238 which remains as 'depleted uranium';
>
> *Fire*: in the heat of the reactor core, the uranium chain-reacts and plutonium is bred; and then
>
> *Water*: in nitric acid baths, the spent reactor fuel is dissolved and thereby the plutonium is separated out - and given to the military.[3]

The 'plutonium economy' is one of stealth and secrecy, as may remind us of the way in which 'Hades was never depicted in ancient Greek art, more out of awe than because of the problems of showing an invisible ruler.'[153]

The problem of who had what plutonium was an exciting government secret, labyrinthine in its deceptions. Pluto's domain gained its 'plutocratic' wealth from minerals, especially precious stones and metals, underground: advocates of a plutonium economy foresaw an era of cheap energy that would ensue from using it. The trouble was that a mere microgram (millionth of a gram) can kill, if lodged in the lungs. The dream/nightmare of plutonium-based reactors, called 'fast-breeder' reactors, died around 'Pluto's turning-point' of 1989, due to world uranium costs dropping.

There is a new prospect for a 'fast-breeder' type reactor called PRISM for the 21st century that will be able to run on all of the plutonium Britain has been producing over the years. This in theory could use up the unnatural man-made element, using it to warm cities slowly rather than ignite them abruptly. It should be taken out of bombs and its energy released, gradually fed into the electricity supply. Yes this will be dangerous but may be the only option to exit from the

---

[153] K.McLeish, 'Myth,' 1996 Bloomsbury p.236.

'plutonium' nightmare. Pluto's unnatural element has to be slowly and gradually released from its existence.

Figure: a Greek alchemical text discusses the *ouroboros* swallowing its tail.

Pluto Rising

What new element before us unborn in nature? Is there a new thing under the Sun?
At last inquisitive Whitman a modern epic, detonative, Scientific theme
First penned unmindful by Doctor Seaborg with poisonous hand,
Named for Death's planet through the sea beyond Uranus...

Ginsberg, Plutonian Ode

There was a definite moment, when the endeavour to create a sample of plutonium began: in Berkeley, California, Glen Seaborg switched on the beam of the big cyclotron onto a sample of uranium.

Seaborg's diaries give us the exact moment of this event. The previous summer neptunium had been made, and Seaborg's team decided to have a go for element 94. A beam of deuterium was focussed upon a uranium sample from 8.00 hours until midnight, on December 14th, 1940.[154] The chart for this moment has Pluto rising within half a degree. Sun, Moon and Earth are aligned (at Full-Moon) with the Galactic Centre at 27° of Sagittarius. That's our local black hole - or, white hole, according to one's theoretical bias - which seems symbolically quite appropriate. Also in line is Seaborg's own Pluto, i.e. its position when he was born, at 27° of Gemini.

The Plutonium-creation chart is bristling with pentagram-symmetries which astrologers call 'quintiles.' There was a Jupiter-Saturn conjunction chiming and Pluto was square to this, and also it was in quintile to (72°) Uranus. That quintile between Pluto and Uranus met and re-met altogether five times (due to the retrograde motions), and Plutonium was created at the *last of these five quintiles*. The Moon was on the midpoint of this aspect i.e. in a decile (36°) aspect to them, and right on the position of Seaborg's natal Pluto (within 5'). Thus Pluto had moved a decile (36°) since Seaborg was born.

Professor Seaborg kindly sent to this writer a copy of his diary for that day which gives the very minute when the cyclotron was switched on for the irradiation experiment, to breed plutonium from a sample of uranium. He has also kindly given his time of birth to Tom Shanks,[155] so we have here an invention or genesis-moment of great value for astrologers, as well as the birthchart of the scientist, which is unusual. The cyclotron was switched on at 8.00 EST, or 4 am GMT on Dec. 15th, then switched off at 12 pm EST.

The chart for plutonium's birth, that is its first creation was during the last phase of a triple Jupiter-Saturn conjunction, with Pluto exactly on the ascendent and various persons have commented that it seems an appropriate chart. There was an exact Full Moon, aligned with the Galactic Centre, and Seaborg's Pluto was conjunct the Moon

---

[154] *The Plutonium Story, The Journals of Professor Glen T. Seaborg 1939-46,* Ohio 1994, p.14. N.K., 'Pluto and Plutonium' *The Astrological Journal* Autumn 1984 p.4. I obtained the relevant page of his then-unpublished diaries via a letter Seaborg kindly sent to me. The genesis-moment is 4.00 am GMT on Dec 15th 1940.

[155] Tom Shanks kindly contacted Seaborg to obtain his hour of birth, which was 4 am on 19.4.1912 at Ishpemming, Michigan (10 am GMT)

to within 5' when he switched on the Berkeley cyclotron to create the first sample. If a small black hole exists at the centre of our galaxy then such an alignment could be symbolically quite appropriate. Pluto remained in square to that Saturn-Jupiter conjunction over the next month or two while the new element was isolated and recognised. Seaborg's Uranus was opposite the Pluto of the creation chart within 1°. His Pluto was in sextile to his Sun at 2° orb, in square to his ascendent at 2° orb and conjunct his I.C. at 2° orb!

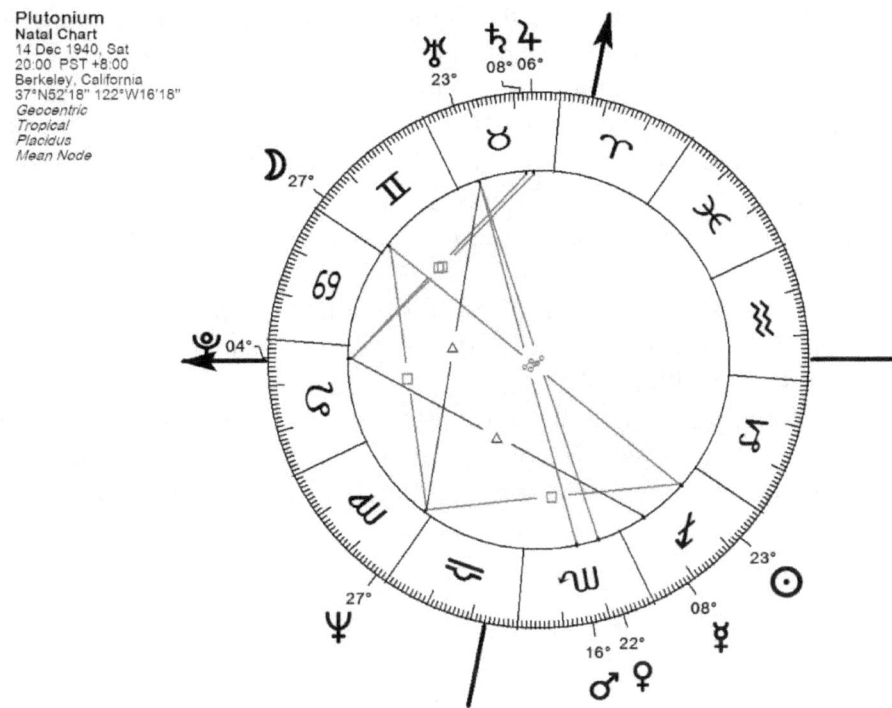

Plutonium has five different crystal-type conditions or 'phases' that it can be in, and has five possible valencies.[156] These are abnormal and surely unique properties for a metal. In addition it seemed to me that it could exist as five possible isotopes that were important, though

---

[156] Plutonium 'undergoes no less than five phase transitions between room temperature and its melting-point.' Also its ions are commonly 'in the III, IV,V & VI oxidation states, but also VII': J.Katz and G.Seaborg, *Chemistry of the Actinide elements* 1957 p.265.

others were also feasible. Its markedly fivefold character was expressed by the pentagram-geometry in the heavens at its birth.

I found and published the chart for plutonium in 1984, then more recently in 2000 its striking connection to the earlier chart for Pluto's appearance was noticed.[157] The 'ascendent' of the plutonium chart was four-and-a-half degrees of Leo, i.e. this was the degree rising when the cyclotron beam was switched on, and Pluto was then at four degrees of Leo. When Pluto was discovered a decade earlier (by Clyde Tombaugh at Flagstaff, Arizona, at 4.00 am on Feb 18th, 1930), the ascendent at Flagstaff was three-and-a-half degrees of Leo: the genesis-moment for plutonium had Pluto rising and on the ascendent degree of its own discovery! Such synchrony rules out the possibility of chance, and indicates that the new metal is in some sense ruled by the new planet. Also, Taylor noticed, a straight line between Berkeley, where the new element was made, and Trinity where the first plutonium device was exploded, passes right through Flagstaff in Arizona.

Allen Ginsberg completed his sombre *Plutonium Ode,* twice-quoted above, on July 14th, 1978. His Pluto-square return was then chiming, to within half a degree.

Pluto and its large moon revolve around a center of gravity which is outside them both, like some biune fission product. Thus the position given for Pluto is actually pure nothingness, mere empty space...

****

---

[157] Brian Taylor, The discovery of Pluto in: *Orpheus, Voices in contemporary Astrology* Ed S.Harvey 2000 247-330.

12. METALLIC MOMENTS

Let us look at some key discovery moments in science, when a specific metal was involved in the experiment or process, and consider the condition of the planet associated with it in the chart. Seven such moments are here presented. We have taken noon times where no record of the timing of events is preserved, though the first two are an exception to this.

The Solenoid: Mars & Venus

'Probably the greatest electrical discovery in history'[6] according to Isaac Asimov was made by Michael Faraday on 17 October, 1831, when he discovered how to make electricity by using a magnet. A cylindrical iron magnet was thrust into a coil of copper wire in an impressive mineral version of the sexual act. Mars is associated by astrologers with the masculine gender, and it is the Mars-metal iron which has the magnetic field in this experiment, so that its movement induced electricity. As it is pulled in and out a pulsating rhythm of alternating current flows through the copper, the pliable Venus-metal, used here because it is a good conductor of electricity. Faraday's diary records how, removing the magnet from the coil, 'A powerful pull whirling the galvo needle round many times was given'. The original solenoid made by Faraday is in the Royal Institution. It is, if you'll pardon the expression, life-size. In the sky on that day was a conjunction of Mars and Venus, a situation which recalls the root meaning of *conjunctio*.

The Sun and Venus stand on either side of Mars, with Venus 6½° away. A Uranus-Jupiter conjunction trines both Mars and Venus, and Pluto has a very close opposition to Venus, within 3'. At 5 p.m. that day the Moon came to oppose Saturn.

Venus was conjunct Faraday's natal Jupiter and Mars was conjunct his natal Saturn, both within a degree, reinforcing the polarity. One may compare this day with Faraday's eureka experience a month earlier, when Faraday 'detected the induced current', on 29 August, again with Venus trine Uranus within 5°. He had a ring of iron a foot across, with two copper wires coiled around it. When a current was passed through one, he found a surge of current appeared in the other. This experimental apparatus was the first transformer.

Entering the Atom

On 22 October, 1934, the Italian physicist Enrico Fermi discovered the secret of entering the atomic nucleus: his team was able to generate *slow neutrons*, and these could penetrate the atomic nucleus when the faster neutrons just passed right by. Silver was the metal used in this experiment. Without going into details, its sensitive and receptive Moon nature responded to the radioactive source and made it the ideal element for this. Also the half-life of the silver isotope formed was short enough for the experiments to be repeated after a few minutes.[158]

Over a 'solitary lunch' the meaning of these experiments dawned upon Fermi - 'the most important discovery I have made' he was to recall - and after explaining it to his colleagues they took their silver cylinder to a local goldfish pond adjacent to their physics department in Rome. If the theory was correct, the water should slow down ('moderate ' is the technical term) the neutrons from the source and the radioactivity induced in the silver should be greatly increased. It worked and everyone became excited. Water is nowadays used in PWR-type nuclear reactors all over the world for just this purpose, but that moment of immersing the silver cylinder in the goldfish pond was the very first application of the principle. Surely, I thought, here is a fine lunar image.

A Full Moon occurred at 4 p.m.! No wonder the team were excited. In fact there were five lunar aspects in the sky. When they came to write a report on the discovery that evening at Eduardo Amaldi's flat, the maid shyly inquired as to whether the guests were tipsy, because of the noise they were making. Uranus was exactly conjunct the Moon that afternoon, indicating the moment in time when Fermi commenced nuclear irradiation of the silver nucleus. On the other side of this mighty opposition, the Sun was conjunct Jupiter and Venus. Pluto is strategically placed, in square to the Sun, Moon

---

[158]. Mike O'Neill wrote a letter in Italian to E. Amaldi to confirm this date, as the several Fermi biographies give different days for it.

and Uranus: this little planet is reputed to be connected with atomic energy. Neptune also forms a fine right angle to the Sun- Moon axis. For scientific work, Saturn seems happily aspected, trine Venus and square Mercury. That afternoon there were altogether five lunar aspects, which must have helped silver to perform its historic role.

The chart is shown for 3 p.m. when Fermi told his team of his new theory, after they had returned from their noon siesta, and when they decided to test it in the goldfish-pond. If we ask, 'Why did the idea come to Fermi at this moment in time, and not anyone else?' then we need to examine the chart's Full Moon axis in relation to his natal chart.

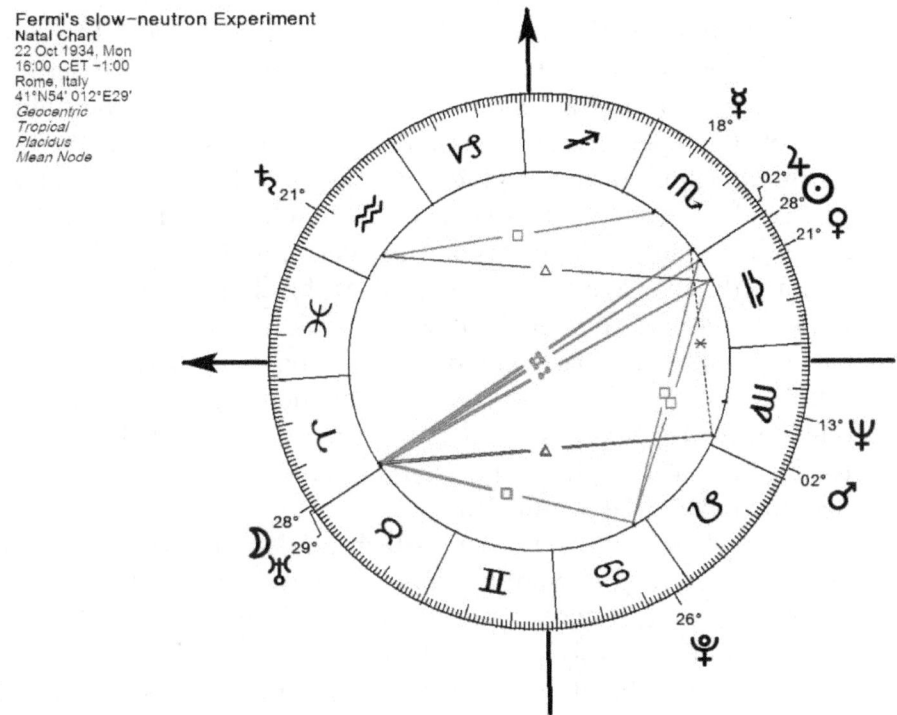

The three outer planets form close aspects to this Sun-Moon axis, in conjunction, square and semisquare. This axis lines up in Fermi's own chart with a Moon-Mercury opposition (MO 27° 1' ARI, ME 28 ° 4' LIB) within a degree. The position of the Sun here, exactly on Fermi's Mercury within ½°, seems most appropriate when the key to entering the atomic nucleus was being discovered. On the other side of the opposition, the Moon exactly conjoins Fermi's natal Moon! There are various other connections with the natal Uranus, such as a close square to Neptune to within a fraction of a degree. Later in the evening, when Fermi's team came to write up their report, the Moon formed a quintile angle (72º) with datum. Quintiles are supposed to be more associated with mental activity, so this seems appropriate for writing a scientific report of an experiment involving a silver apparatus.

Nuclear Power

Fermi's discovery came to its fulfilment on 2 December, 1942, in Chicago, as a part of the wartime Manhatten Project: his team switched on the world's first nuclear reactor. 'The Greatest Experiment of All Time' was how the *Bulletin of Atomic Scientists* described it. Starting in the morning, all day the team slowly withdrew the cadmium control rods from a uranium-graphite pile, and listened as the geiger-counter clicking slowly increased. At 3.30 p.m. Fermi said, 'The chain reaction has begun', as the dials started to increase of their own accord, as a self-sustaining chain reaction got going, and then at 3.53 p.m.[159] Fermi ordered that it be switched off. At the climax of the experiment when, if left on any longer it would have started to overheat and become unduly radioactive, Uranus rose above the horizon and Pluto simultaneously passed the IC. All day the neutron flux had been building up to this maximum point, then it was shut down. This is reminiscent of the plutonium chart, where Pluto was just rising; but here the metal used is uranium.

In the chart for this moment, which marks *the actual beginning of atomic energy,* the Sun conjoins Antares and Saturn conjoins Aldebaran. These two first-magnitude stars are exactly 180º apart, within minutes, and form the principal reference axis for the Sidereal zodiac, being defined as 15º Scorpio and 15º Taurus respectively.[160] Antares, which the Sun conjoins, is a binary star of fiery red and

---

[159] Chicago was then on War Time, which was five hours behind GMT, while the top-secret Manhattan Project remained on Standard Time, which was six hours behind. See E. Troinski's *Das Horoskop des Atom-Zietalters,* p.92-3.(1975, Munich). This was pointed out to me by US astrologer Doris Heber, whose father worked in the Manhattan Project.

[160] This is speculative, argued by siderealists Cyril Fagan and Robert Powell, but Chaldean tablets do not refer to such an axis.

emerald green hues, the 'heart of the scorpion', associated with 'danger of fatality', according to Robson's Fixed Stars and Constellations, while the pale rose star Aldebaran, which is rising, has a better reputation. The midpoint of the Sun-Saturn opposition squares within 5' the two fixed star positions.

A Sun-Uranus opposition has just occurred. As I see it, this chart shows a mighty struggle between Saturn and Uranus, where Saturn represents the fixity and limit of matter, the indivisibility of the atom as it existed for thousands of years, the Saturn-metal lead being the heaviest non-radioactive metal, i.e. whose nucleus is stable: any bigger and the nucleus starts to fall apart. Whether Uranus is linked specifically to uranium I do not know; evidently it was named after the new planet, and the symbolism seems highly appropriate.

A powerful opposition is taking place, with the Sun conjunct Mercury and Venus opposing Saturn, and the latter is strengthened by a conjunction with Uranus and a trine, to less than 1º, to a conjunction of the Moon and Neptune, within 26'. Neptunium would have been produced in the atomic pile, which then decayed into plutonium.

The outer planets come together in a grand trine in this chart, as the MC trines Neptune which trines Uranus, and Pluto at the IC trines the Sun and Mercury. The exact timing for these events was as follows: at 3.20 p.m. the MC trines the Moon conjunct Neptune; at 3.25 the MC trines Uranus, as the first-ever chain reaction begins. At 3.30 Fermi says, 'The chain reaction has begun', and a couple of persons exit to make the phone call saying, 'The Italian navigator has entered the New World ', the code phrase for the Manhatten Project. At 3.45 the Ascendant trines the Moon conjunct Neptune, at 3.50 Uranus and Pluto cross the Ascendant and IC respectively and a few minutes later the reaction is switched off by dropping in the control rods.

A critical part of the chart for this event falls where the Sun, Mercury and Venus all conjoin in opposition to Saturn, just at the point where Fermi's natal Uranus is located. As for the previous experiment involving silver the main contact was with his natal Moon, so here with a uranium experiment the main contact is through Uranus. However there was not such a strong tie-up to his natal chart for this event as for the previous silver experiment, as his contribution was less distinctive and more as part of a team.

Secrets of the Seven Metals

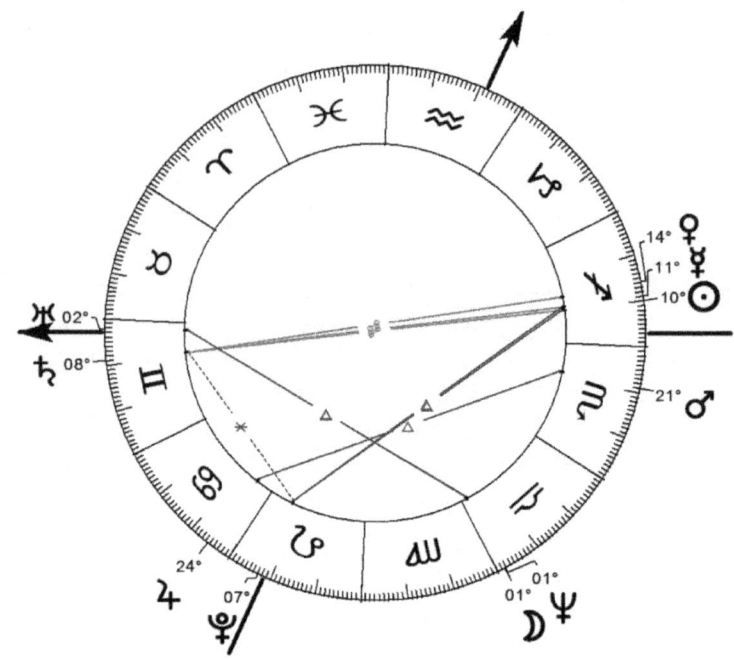

Nuclear power
Natal Chart
2 Dec 1942, Wed
15:50 CST +6:00
Chicago, USA
41°N56' 087°W37'
Geocentric
Tropical
Placidus
Mean Node

    This chart really signifies the Dawn of the Nuclear Age and as such is highly significant. It has several levels of timing: the grand opposition aligning with the two fixed stars Aldebaran and Antares, two stars traditionally linked with war and fiery energy, defining the day of the event; the conjunction of the Moon with Neptune, giving the hour, and finally the alignment of the planets with the angles giving the very minutes of the reactor going critical. Then, had the control rods not dropped, we could, as they say, have lost Chicago.

    There was a later date when a uranium chain reaction went to completion for the first time ever, and that was over the city of Hiroshima. (The earlier bomb at Alamagordo was made from plutonium). Do we expect a relationship between the two charts? The Hiroshima Mars (on August 6th, 1945) was exactly conjunct the axis of the nuclear- energy chart, being within 14' of its Saturn, and the Pluto of Hiroshima has a close link with the 'nuclear axis', being in trine (7' orb) to its Sun. Thus the shadow of Hiroshima does seem to lie on this chart. Fermi was just experiencing his own Mars-return within a degree, so altogether we are left in no doubt as to the military context within which atomic energy was born. Uranus culminates at the MC of the horoscope for the dropping of the atom bomb on Hiroshima.

U.S. astrologer Mark Lerner has argued that the Aldebaran-Antares 'nuclear axis' is relevant to major nuclear disasters. Thus the Chernobyl reactor blew its top on 26 April 1986, as Saturn returned to this axis at 8º 35' of Sagittarius, whereas it had been at 8º 57' of Gemini (i.e., opposite) in the nuclear power chart. It was one degree from Antares 'heart of the Scorpion' and the Sun was opposite Pluto on that day. Going back thirty years i.e. one Saturn-cycle took Mr Lerner to the Windscale nuclear accident, on 7th October 1957, with Saturn at 10º 5' Sagittarius, conjunct the sun of the nuclear power chart.

Mars and the Jet Fighter Plane

Sir Frank Whittle was the 'inventor of the jet engine', according to the Biographical Dictionary of Scientists: 'The developments from his (Whittle's) original designs were used first in the Gloster Meteor fighter at the end of World War II. Direct descendants of these engines are now the sources of power for all kinds of military and civil aircraft.'[161]

The birth of the first British jet plane has been described by John Golley in his book, *Whittle, the True Story*. Fortunately, a chapter entitled, 'The First Flight - 15 May, 1941' locates the time as well as the date of this historic event. The new, jet-propelled aeroplane soared off into the clouds shortly after 7.40 p.m. (17.40 GMT). It resumed 17 minutes later, before a small, amazed group of spectators at Lutterworth airport. One of them clapped Frank Whittle, the inventor of the jet plane, on the back and exclaimed, 'Frank, it flies!' and he is said to have replied, 'That's what it's designed to do, isn't it?' The speed of the Gloster-Meteor on that maiden flight exceeded that of the Spitfire, the plane with which Britain was then fighting World War II: a historic moment for British aviation. So unexpected was the success of this maiden flight that no MOD official had come to film the event. No celebration party had even been planned for that evening. A month earlier, in April, the plane had run on 'taxi flights' when it had

---

[161] Whittle Jet: D.Abbott Ed., *Biographical Dictionary of Scientists, Engineers and Inventors*, 1985, p.156.

just skimmed along the surface of the ground.

May 15 at 17.40 G55MT was therefore the genesis moment for this fiery new iron/Mars instrument. There were at that time no less than *four* square aspects to Mars then present (within 5° of orb): Mars was square to the Sun, (5° orb), square to Jupiter (2°), square Uranus (3°) and square Venus (2°), also square Mercury at 6° of orb. Uranus was conjunct to the Sun, Jupiter, Venus and Saturn! Altogether there were seven major Uranus aspects to 5' orb. A grand trine was present, perhaps helping the moment to be a complete success.

Figure: Whittle's Jet plane takes off, with Mars-aspects.

When did the first locomotive move upon rails? It was built by a Cornishman, Richard Trevithic, in 1804, in response to a wager to haul ten tons of iron from Penydaren to Abercom Wharf by steam power. A letter of his recalled how

> Last Saturday (11 February) we lighted the fire in the Tram Waggon and work'd it without wheels to try the engine; on Monday we put it on the Tram Road.[162]

One would expect Mars to be in strong and sturdy aspects for such a day: it was conjunct Mercury to 4°, in square to Jupiter and trine to Saturn. A conjunction with Mercury, swift patron of travellers, seems appropriate.

The First Computer Program

Appropriately enough, the first computer designs in the 1940s used the metal mercury to store information. The problem was how to have information in a form that was stored yet readily accessible. 'Mercury delay tubes' were used, which were some five metres long and pulses travelled back and forth along them, encoding information. As a recent book on the development of computers commented about this mercurial device:

This appropriate element, associated with the classical deity of speed and communication, was to haunt the developments of the next few years.[163]

Then, on 21 June, 1948, at Manchester University, a team of researchers 'successfully ran the first program on the first working stored program electronic digital computer in the world'.[164]

Earlier dates sometimes cited are for mere calculating machines. I contacted one of the people involved at Manchester and he informed me that the initial program was run at noon, just before lunch. Uranus was then conjunct the Sun, and again the chart shows an exact Full Moon. Uranus is half a degree from the Galactic Centre and 5' from the Sun-Jupiter midpoint. Mercury is at the hub of things, in a wide conjunction (8°) with Uranus so that the midpoint is right over the Sun-Moon axis. It has a strong semisquare to Saturn (16') and a close quintile to Mars (10'), the quintile aspect being associated with creative mental activity. It is within minutes of Venus, quite appropriate considering all the electricity conducted by copper wire on this historic day.

---

[162] L.Rolt, *The Cornish Giant*, 1960, p.81.
[163] A. Hodges, Alan Turing, *The Enigma of Intelligence,* 1984, p.315.
[164] Ibid, p.385.

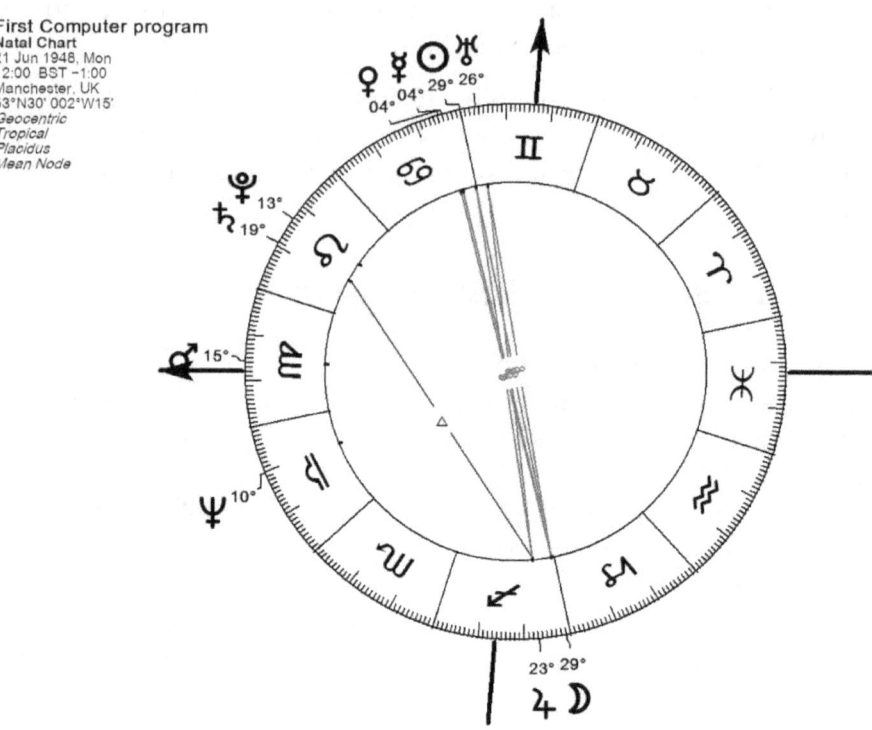

The scientist whose name is primarily linked to the construction of this early Manchester computer is F. C. Williams. Taking a noon time for his birth on 26 June, 1911, when a New Moon occurred, his Sun falls in exact conjunction to the Mercury of the computer chart within 15'. So the natal Sun and Moon fall right on the Sun-Moon axis of the computer chart, and at a wider orb also conjoin Venus and Uranus. The natal Mercury conjoins the Uranus of the computer chart within 2', and his natal Pluto within 1°. So altogether we find a remarkable concordance of the two charts.

The Superconductor and Venus

When the first superconductor worked it caused a sensation in the scientific world. Chinese physicist Ching-Wu Chu working at Houston, Texas was awarded the Nobel Prize for this achievement.

Metallic moments

To quote from a recent book on the subject:[165]

> In January 1987 Professor Paul Chu of the University of Houston made a major scientific breakthrough that was to reverberate around the world.
>
> For many years superconductors, materials that conduct electricity with virtually no resistance, have been known about and available for commercial use -- but at operating temperatures too low to make their wider application economic. Paul Chu's achievement was to shatter all previous rec ords and produce a superconduc tor at workable temperatures.. . His discovery is to physics what the double helix was to biology and its full implications are yet to be realised.

When Chu mixed three elements, barium, yttrium and copper, he committed alchemy.

On the 29th of that month, 'At 5.00 p.m. the measurement was made... All of the Alabama and Houston workers were exultant. It was the discovery of a lifetime '. The laboratory was at Houston, making the time 23.00 hours GMT.

Superconductors are special ceramics made of copper complexes. The one made by Chu was of a green hue, copper's colour. Copper displays the Venus-property of conductivity in copper wires to conduct electricity for example, or in solar panels on a roof to collect heat (which are normally copper-based). At Houston this property appeared in a most remarkable form. At the historic moment there were no less than five major Venus aspects taking place: it was conjunct Uranus (2º), conjunct Saturn (5º), trine MC (3º) square Jupiter (20') and sextile Mercury (2º). It was a moment when electricity somehow overcame the inertia of matter, so that it could glide past the copper atoms without resistance: for which Venus conjunct both Uranus and datum and textile to Mercury seems most appropriate. Uranus, the planet which astrologers associate with electricity, was also conjunct Saturn (5º), and square to Jupiter (2º).

---

[165] R. M. Hazen, *Superconductors: The Breakthrough*, 1988, p.52.

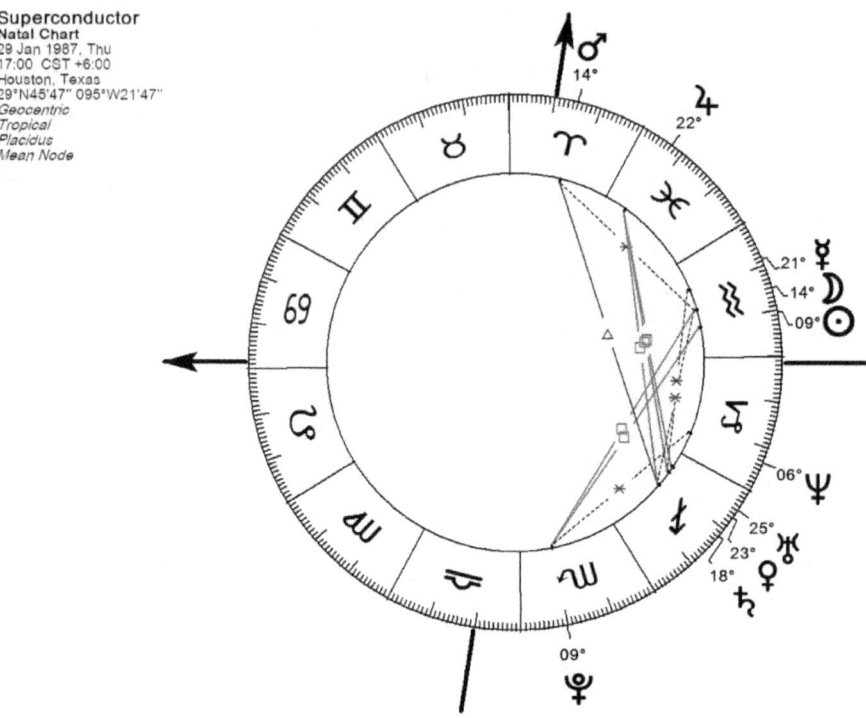

Superconductor
Natal Chart
29 Jan 1987, Thu
17:00 CST +6:00
Houston, Texas
29°N45'47" 095°W21'47"
Geocentric
Tropical
Placidus
Mean Node

It was a New Moon, with the Moon in conjunction with both Sun and Mercury to around 6° orb, i.e. the Moon was near the midpoint of these two. As Mercury with the two luminaries moved down to touch the horizon, a new property was born in matter. The New Moon was chiming an exact septile to Venus (16'), as too were the Sun and Saturn. Three septiles were present, making the moment quite an inspirational one.[166]

In the moments described above, the *qualities* of the metals find expression. Thus silver the Moon- metal is of a sensitive and receptive nature, and it was just the metal to detect the slowing-down of neutrons which Fermi's team managed to accomplish. It absorbed the slow neutrons from a radium source and so gave the key to entering the atom. Again, mercury's nature pertains to transformation, and this is how oxygen was discovered: on heating mercury, it will absorb oxygen, and then on further heating as Priestley did, it gives it off again.

---

[166] See NK, *Eureka the Celestial Pattern at Moments of Historic Inspiration,* 2012.

Uranus & Neptune Transits

We've looked at two uranium charts, nuclear fission in 1942 at Chicago and the atom bomb at Hiroshima four years later. In both cases Uranus was 'angular', as an astrologer would say, i.e. it was on the ascendant and then the MC, respectively. Both charts strangely gave the impression of being timed by this event, as if Fermi had to wait until Uranus-conjunct-Saturn rose above the horizon, before he could smile and say 'the chain reaction is accomplished' - when the reporter rushed to the telephone and spoke the code words, 'The Italian navigator has entered the New World'; somewhat as if the crew of the Enola Gay had to wait until the very minute of Uranus' arrival at the local MC before releasing their dreadful cargo. Those two events were, to remind the reader, the first controlled fission of uranium, the chart for the 'dawn of the nuclear age', and the first uncontrolled fission of uranium, i.e. a bomb. They were linked to each other, in a powerful synastry between the two charts.

Neptune was discovered during its conjunction with Saturn (within 45 minutes of arc) in 1846. There had been claims about a new planet for a year or so, but only after that conjunction had struck did an astronomer look up in the sky and see it, on September 23, 1846. The two mathematicians concerned, Leverrier and Adams, French and English, came up with the same discovery quite independently, no doubt helped by a Neptune-Saturn conjunction in their synastry (i.e., between their two charts) within one degree. This expressed what they shared in common. That discovery consolidated existing scientific law, in this case the theory of gravity, and so the Saturn conjunction was appropriate. The conjunction was just passing over Leverrier's natal Mercury when he sent off a historic letter (September 18th) to the Berlin observatory telling them where to look for Neptune. That transit was symbolically appropriate for the sending of such a letter. What the scientist discovered in the heavens was working also within his own psyche, for we are all a part of the world-process.

Meanwhile, in America, a chemical invention was taking place, as the new sphere appeared in the depths of space. A Boston dentist William Morton was purifying a sample of ether, taking advice from his teacher Charles Jackson on how to do this. On September 30th of 1846, when the Saturn-Neptune conjunction was exact within one degree, Morton successfully sent himself into unconsciousness using his ether sample. He was then 27 years old, with his Neptune sextile-return within 16' on this date (i.e. 60º from its position at his birth). Two weeks later, on 16th October, Morton administered the first ether anaesthesia used in a surgical operation, at Massachusetts General Hospital, and history was made.

The surgeon's operating room and the torture-chamber finally parted company, as the wondrous new unconsciousness-without-pain arrived, a present from Neptune. Prior to Morton, doctors had regarded ether as a deadly poison, and many continued to do so. An astrologer would point out that a conjunction of Saturn with Neptune well symbolizes the concept of ether being purified. Morton's natal chart has quite strong synastry with the anaesthesia chart, via Neptune. Morton and Jackson had a dire lifetime feud over priority claims, and only posthumously did Morton receive his credit. The whole question of who made the discovery was for long wrapped in a miasma of confusion, which as an astrologer would point out is very Neptunian. Jackson's Neptune formed a semisextile (30') to Morton's within 15 minutes of arc, so they were both experiencing Neptune transits during the great discovery (of 60º and 90º, respectively) - their controversy was 'written in the stars'. At this point one experiences frustration that, in such a classic textbook case, both times of birth have been lost.

The Red Sands of Mars

He learned chemistry, that starry science...
                    Moffat's biography of Sir Philip Sidney[167]

RED STORMS RAGE across Mars: iron-bearing dust particles on Mars swirl up in giant storms lasting for weeks which blot out all of its surface features. In 1976, Carl Sagan and the team at Pasadena controlled the descent of the Viking spacecraft onto the sands of Mars, and found it an enthralling but perilous experience.[168] Previous attempts by the Soviets to soft land on Mars had been thwarted by the violence of the Martian climate, and their craft had been smashed to

---
[167] Quoted in C. Nicholl, *'The Chemical Theatre'*, London: Routledge & Kegan Paul, 1980 p.15
[168] C.Sagan, *'Cosmos'*, London, Macdonald, 1980 Ch.V.

pieces. Slowly, through the pink skies, an iron instrument descended onto the sands of Mars.

Carl Sagan's book, *Cosmos* describes how the event had a major importance in his life. For an astrologer such a pivotal moment has to be linked to the natal chart of Sagan. Iron/Mars has to have a role in this linkup to the natal chart - in accord with the words of the astronomer Johannes Kepler:

> And since almost every motion of the body or soul or its transition to a new state occurs at a moment when the figure of the heavens corresponds to its birth figure (which is usually only a matter of certain correspondences in detail), it happens that some notable men will be most greatly moved by these aspects...[169]

Just for the record, I wrote the above paragraph *before* finding the transits! Mars was conjunct Sagan's natal Mars, at $4°$ orb, on that historic morning of July 20th, 1976, at sunrise at the Jet Propulsion laboratory at Pasadena, when the first spacecraft was landed on another planet.[170] Also, his Uranus was in opposition to Uranus to $4°$ orb. As one who did much to criticise the 'irrational' tenets of astrology, one regrets that this great modern astronomer did not ever ponder the significance of the Mars-return plus his Uranus-return, at this climactic moment of his life. At touchdown, a moment of triumph and celebration after a sleepless night, the Sun touched the horizon at Pasadena as Jupiter touched the ascendant of Carl Sagan.

A few years earlier in 1971, when the Mariner 9 spacecraft was sent into orbit around Mars, and found it wholly covered by a duststorm, the theory of nuclear winter dawned upon Carl Sagan, concerning the unwinnability of nuclear war.[171] There was not much that the Viking spaceprobe could do while the iron-duststorm obliterated the Sun's light from the surface of Mars, and during those days Sagan pondered

---

[169] J.Kepler, '*De Fundamentis Astrologiae Certioribus,* Prague 1602 (English translation by Bruce Brackenridge & Mary Rossi is entitled, 'On The More Certain Fundamentals of Astrology', in *Proceedings of the American Philosophical Association* 1979; also, J.V.Field, 'A Lutheran Astrologer: Johannes Kepler, in *Archives for History of Exact Sciences,* 1984, 31, p.268.)

[170] 'Viking, The Exploration of Mars' NASA publication, 1984, Jet Propulsion Laboratory, Pasadena, p.3.

[171] P.Erlich et. al., *'The Nuclear Winter',* London: Sidgwick & Jackson,1985, p.1.

the lowering of surface temperature as a result of the darkness. The terrible implications of nuclear weapons dropped onto cities on Earth dawned upon him; global nuclear strategies have altered from this insight, this new message from Mars!

For his leading role in the Viking and Mariner expeditions, Carl Sagan was awarded NASA medals and also gained an international astronautics prize. Transits help us to look at the subjective involvement of the scientist in his experiment, which is something that always gets excluded from the final report. Is not the Mars return to Sagan's chart just as precise and just as interesting as the almost twenty percent of iron which the Viking measured in the surface soil?

The study of such transits can help us to appreciate how astrology works - as studying the characters of the seven metals helps us to experience the reality of its archetypes.

'The soul of hardened steel...'

Iron/Mars is bold and direct in its symbolism and mode of expression. In his 'Astrology Disproved', the US sceptic F. Jerome claimed that astrological beliefs had a primitive, prescientific logic - for example that Mars was red and *therefore* had become linked to blood and war. That is a fair statement, *provided* one appreciates the metallic connection: iron, the Mars-metal, has turned the whole surface of Mars a rusty red hue - indicating that oxygen was once present there - and is the metal in the blood which makes it red, as it carries the oxygen towards the fire-processes in the human organism; and is also the element of which instruments of war are made.

Michel Gauquelin found that the two professions of soldiers and sports champions were linked to Mars by their birthtimes. He concluded that they had in common the 'Iron-willed temperament', a temperament prone to appear in persons of eminence born when Mars was rising or culminating. A fine phrase he used for this temperament was 'the soul of hardened steel characteristic of the true sportsman.'[172] Great storms of controversy swirled around his claim to have 'proved' his planetary-destiny effect with eminent persons, especially famous sportsmen, ending with his suicide in 1991. This

---

[172] Gauquelin,M. *'Cosmic Influences on Human behaviour'*, London: Garnstone Press, 1977, p.100.

was a sad climax to his life's work. His Mars-effect was valid, I have argued,[173] he just got a bit too personal about it.

In 1988 two Americans Hill and Thompson reported that a sample of 500 redheads tended to have Mars near to the ascendent more often than would be expected by chance, claiming their result was highly significant.[174] They observed an excess of Mars positions within 30 of the ascendent, and a deficit within 30 of the descendent. Since time immemorial, astrologers have claimed that Mars types have red hair. Red colour in hair is produced by the iron pigment trichosiderin, and it is genetically determined. It would be startling if so concrete a demonstration of planetary influence could indeed be demonstrated. To establish a degree of planetary determinism on a trait regarded as genetically coded would have far-reaching repercussions. The two women became rather burnt-out by the strong controversy which their results generated.

This chapter has been developed from a talk given at the Astrological Lodge on 15 June, 1987. The talk discussed the Sun-Uranus opposition present in the 'atomic power' chart, and how it was aligned with these same two planets in Fermi's nativity: how the Sun was conjunct Uranus and Uranus was trine his Sun. It happened to coincide with the same opposition aspect present in the sky on that day aligned with a Sun-Uranus opposition in my own chart, along its meridian. Was cosmos trying to tell me something? The talk explored the nature of planet-metal affinities in moments of scientific discovery: it had not at that time dawned upon me that Uranus was a key factor in these invention-moments.

---

[173] N.K., 'How Ertel Rescued the Gauquelin Effect', *Correlation* 2005 (online)
[174] J.Hill and J.Thompson, 'The Mars-Redhead Link' *NCGR Journal,* Winter 1988/9 (also The Astrological Journal, September 1988).

EPILOGUE

As I have already said, the term imaginatio, like meditatio, is of particular importance in the alchemical opus.
                    Carl Gustav Jung, *Psychology and Alchemy*, p.276.

Her Heart of Fire

So what is alchemy? That is hard to say, but it might help to study an image by the great alchemist Michael Maier (1588-1622). He lived in Prague, centre of European alchemy, and was physician to the Emperor. This image gives us meaning without words. I call it, 'Her Heart of Fire.' A young man has been rebuffed, he has dreamed of a woman's body, shown centre, and his Mars fire-energy now burns with shame and futility. With one hand he holds a hammer, so maybe he is a smith. She is manifesting the fire of her heart. Her heart burns. He has failed to perceive this, he desired her body.

Epilogue

Figure: from *Symbola aurea mensae*, Frankfurt, 1617 Michael Maier

The sage holds his book of wisdom, and remains calm. Winds blow the woman's hair and cloak around, signifying stormy passion, yet his garb remains unruffled. He is pointing to her heart, to her heart of fire. The thwarted Mars-figure in the rear has the line of his back crossing her genitals. Two different kinds of fire-energy are here depicted, that of *Mars* and that of *Sol*. She represents *Natura* or Mother Nature, and the sage advises us to cognize her inner being, her heart, not merely her outer glory. This is a warning to future centuries who would view Nature as a body without soul, to be dissected. Maier's book title translates as, *Golden Symbols of the Mind.*

A century earlier, an alchemist from Frankfurt wrote, with his quill pen:

> And let thy imagination be guided wholly by nature... And imagine this with a true and not with a fantastical imagination.
> *Rosarium,* Frankfurt 1550

Jung attached importance to this passage, and he explained, 'The *imaginatio* is to be understood here as the real and literal power to create images, the classical use of the word in contrast to *phantasia* which mean a mere 'conceit' in the sense of insubstantial thought. ...This activity is an opus, a work.' (bid, p167[175]) That advice may seem odd to us given the fantastic nature of alchemic images, but we have here endeavored to follow it.

Following on the pioneer work of Kolisko, several books were published on the metal/ planet theme in the last half-century by Rudolf Hauschka, Wilhelm Pelikan, Mellie Uyldert and Alison Davidson (in the Foreword we showed the front covers) each with a distinctive emphasis. Each of them found a *larger number* of correspondences with modem elements than has here been accepted, making the present work appear as fairly conservative.

Rudolf Hauschka's *The Nature of Substance* diagrammmed a spiral of creation whereby '...oxygen comes into being under the

---

[175] I sort of agree with Adam McLean who wrote to me, "Jung twists the meaning by missing out the interlinking passage, and removes the context, in order to give it a psychological spin." But, that is what Jung wrote.

influence of Aquarius in the reflected Mercury-sphere...' Taking the above advice to imagine 'with a true and not a fantastical imagination,' we may view this as the product of an unduly vivid imagination. It is *too extravagant*. Some elderly readers may here be reminded of Rodney Collins' *The theory of Celestial Influence* which likewise had a tendency to dream up complicated webs of correspondences, then believe in them.

Planetary Images

The planet Venus weaves out a pattern of perfect beauty, harmony and proportion. My book *Venus the Path of Beauty* explored this theme. This may help us with the concept of *signature* as Paracelsus used it: is the universe set up so that *archetypal being* is expressed in *signatures* upon metals and planets?

Figure: Plotting Venus' motion with respect to Earth at the centre, over its quantum eight-year period.

The first diagram shows the orbit of Venus in sidereal space viewed geocentrically over its eight-year period (as it were, in polar co-ordinates), while the second plots straight-line vectors connecting Earth and Venus, every two days of so, from heliocentric co-ordinates, over the same period. The second image emphasizes more the

Epilogue

concept of a relationship between two spheres, woven in space. We are puzzled, why these two figures should look so similar.

Figure: straight lines join Earth and Venus, as they go around the Sun

This pattern should be a part of the education of every schoolchild in their mathematics lesson, just as the imaginative perceptions discussed in this work should be used to imbue the chemistry lessons with meaning and delight. The rose-pattern, with its five 'hearts' woven every eight years, revolving so that the same portion of the planet comes to face earthwards at each inferior conjunction (i.e., when it is nearest to the Earth) is a reassuring image of harmony and beauty, woven around the earth.[176] It is apparently without interest to astronomers, not being found in their textbooks, because the pattern only appears from a geocentric reference. That does not concern them, they have as it were left it behind.

Harmony and beauty are seen in the Venus-pattern, being the *signature* of the power involved. That is an astrological statement rather than an astronomical one, since it contains a value judgement about quality. For the astronomer, the phenomenon merely arises because the two orbits (Earth and Venus) happen to be closely circular, and he surmises that there is some resonance effect causing

---

[176] NK, *Venus the Path of Beauty* 2012 (Venus the Heart and the Rose, online).

the whole pattern to be formed every eight years to the day -though we lack any evidence for such resonance.

In contrast, rough Mars has no exquisite proportions as does Venus, in its motion. It is very asymmetric, with a large eccentricity. Venus forms the golden ratio in its orbit period in relation to Earth's year, but Mars does nothing like that. However, let's notice that the two have a lovely 3:4 rhythm in their approaches to Earth. Venus comes closest to Earth in its inferior conjunction, when its invisible for a week or so we cannot see it. After it fades away from our view into the sunset as the Evening star, then a few days later it comes closest to earth, moving retrograde. Mars in contrast is very visible glowering brightly in the night sky when it comes closest to us, every two years or so.

Comparing these intervals of coming-closest, *three Mars-periods equal four Venus-periods,* to within 99.8%[177] Thjat's pretty close! These are the synodic periods of the two planets. This is like, a waltz-rhythm dance of movement. The synodic cycles of Mars and Venus are the motions as we experience them, as we see them in the night sky. This harmony appears *in relation to* Earth, our home, where we are - so therefore, astronomers aren't interested.

Let's continue to look at signatures in modern planetary images. The Viking spacecraft discovered that continual cascades of lightning took place in the atmosphere of the planet Jupiter, far larger than lightnings on Earth, and that these were the cause of the brilliant colouration of that giant orb. Such discharges are linked to the vast magnetic field of Jupiter, whose magnetic storms reverberate across the solar system. This astronomical fact establishes a connection between present-day astronomy and the ancient mythic figures of Jove or Zeus - and of Thor, the Norse thunder-god (Thursday = Thor's day, as Jeudi = Jupiter's day).

Astrophysicists may be able to account for why Jupiter alone amongst the planets has continuous lightning in its atmosphere, or they may not, but this hardly concerns us. What matters here, is that what Paracelsus called the 'signature' of the archetype (or whatever one calls it) is here manifesting. The lightning is an expression of the inner being, the principle that we call Jupiter or Jove.

---

[177] Mars' mean synodic period 779.7 days, Venus 583.9 days: discovered by John Martineau, in *A Little Book of Coincidence,* read it in: www.woodenbooks.com/.

When that same spacecraft arrived at Saturn, it 'saw ' a completely different picture. The *National Geographic* reporter described it as follows:

> The pictures coming across the monitors speak directly to the imagination. Not fiery, chaotic and psychedelic like Jupiter, they look cool, ethereal, and from a distance orderly enough to have been drawn with a celestial compass.[178]

These are profound matters, and I suspect it will be a long time before the science of today becomes capable of appreciating them: a new generation will have to arise, capable of using their imaginations instructively and restricting the abstracting intellect to the domain in which it belongs. When that day comes, when the abyss between astrologers and astronomers has been apprehended as mere deficiency of perception, then the extraordinary symmetry and precision of those level rings around datum will be an uplifting source of inspiration. It will be apprehended that we inhabit a living universe, not a dead one.

Signature and Rulerships

Astrology uses the traditional notion of 'rulership '. This concept is similar to the ideas of 'affinity ' and 'signature ' employed in the present work, though it is more heirarchical. An analogy with traditional plant rulerships is here useful. A one-to-one correspondence between the seven metals and their respective planets has here been advocated, but does any comparable notion apply to plants? A *Book of Rulerships* by US astrologer Dr Lee Lehman compared traditional rulerships in a scholarly manner, showing how the rose for example was generally assigned to Venus by astrologers of old, but with an exception: the Arab Al-Biruni claimed that it was ruled by Mars.

Kolisko 's work, *Agriculture of Tomorrow*, discussed a few rulerships as given by Rudolf Steiner which differed from tradition, e.g. the oak tree with Mars - traditionally the oak was always linked to Jupiter. In chapter five, while discussing the work of Agnes Fyfe, we alluded to the traditional concept that mistletoe had a solar virtue. From the standpoint of modem botany the argument appears

---

[178] 'Voyager 1 at Saturn', *National Geographic*, July 1981.

nonsensical, but that is not the view here taken: the botanist has thrown out the baby with the bathwater, has lost the kernel of truth, dealing merely with the outer husks of appearance, and has failed to make a proper use of the imagination whereby these things may be discerned.

Dr Lehman's book gave a table of rulerships where her historical sources all agreed, or almost all agreed: for example that the apple was ruled by Venus, the onion by Mars, hemlock by datum and vines by the Sun. The implication here is that one should wait for a good Sun-Jupiter aspect (for example) before planting a vine. On the other hand, this list contained no 'Mercurial ' plants. Her comment was,

> You may notice in examining the Table that there are no Mercury- ruled plants listed. All the authors assigned plants to Mercury, but multiple authors did not agree on Mercury rulership for the same plants. This is especially interesting because all authors stated that Mercury takes on the colouration of other planets with which it associates. 'Mixed ' and 'complex ' were frequently used words to describe Mercury. It thus appears that quicksilver eludes even the astrologers! One may also note, that a rose, by any other name may smell as sweet, but if. like Al-Btruni, you are concerned with thorns, then it is Mars-ruled!

The signature of Mercury here manifests, in the inability of astrologers to fix its rulerships!

Let's conclude with an ancient Chinese statement about alchemy, from the 2nd century BC, an early formulation of the subject. It could be spoken by people standing in a circle, each one reading a line:

I met the Feathered Ones at Cinnabar Hill

I tarried in the ancient Land of Deathlessness

In the morning I washed my hair in the Hot Springs of Sunshine

In the evening I dried myself where the ten suns perch.

I sipped the subtle potion of the Flying Springs,

And held in my bosom the radiant metallous jade.

My pallid countenance flushed with brilliant colour.

Purified, my ching of vitality began to grow stronger;
My corporeal parts dissolved to a soft suppleness,
And my spirit grew lissom and eager for movement.
I attained the Clarity and entered the precincts
Of the Great Beginning.[179]

★★★★★★★★★★★★★★★★

---

[179] Yuan Yu, 110 BC: Peter Marshall, *The Philosopher's Stone*, 2001, p.34.

APPENDICES

1. The Great British lead-debate

'Gasoline lead may explain as much as .90 percent of the rise and fall of violent crime over the past half century'

– Mother Jones report.

The University of Surrey's chemistry department is proud of having quite a high-powered atomic absorption spectrophotometer system. That apparatus measures matter as light. It measures the presence of atoms at high-temparature, in other words it will only measure elements, not compounds. It will read out the concentrations of a whole string of metals from a single sample. And – my colleague Dr Neil Ward proudly advised me - it measures in parts per billion. At last science has developed to be able to measure the toxic heavy metals at the low levels where they start to injure the growing central nervoussystem.

Just think about it – Britons have been brain-damaging themselves with lead for thousands of years or at least since the Romans came, with pipes, pewter, paint, car exhaust etc., and now finally we are able to measure reliably down to the toxic threshold that really matters..

And what is that toxic threshold? Well, its somewhere around one-tenth of a part per million in human blood. Today lead still remains the one clear pollutant where the ambient level in blood overlaps the threshold at which toxic effects begin to show up. And we are here talking about a five-point or so reduction in human intelligence, in I.Q. Or at least that's what my book *Lead on the Brain* argued way back in 1982.

Some readers may fondly imagine that there exists some kind of government environmental protection agency which is concerned to monitor these things, and would utter a warning if there was reason to suspect that a generation of heavy metal kids were having their central nervous systems damaged by toxic metals. Alas that is far from being the case. We do have some EU guidelines given of maximal permitted levels of pollutants, that is about the nearest there is to any such thing.

The public have very little interest in the 'science-fiction' idea that low 'homeopathic' levels such as one hundred micrograms per kilogram (that's D7, or what I'm here calling one-tenth of a a part per million) of lead or aluminium may be doing anything. Receiving funding from big business and wishing to support industry, the government is not in a hurry to legislate anything that will impede their productivity.

But, let's suppose that a green, eco-friendly government appeared that was concerned about irreversible stunting of child IQ and the concomitant rise in violent crime which results from a witches'brew of toxic heavy meatal a-jangling in the bloodstream of inner-city youngsters. Hyperactivity is a condition greatly increasing nowadays and it is also closely linked to what we are here discussing: this is the child who cannot sit still in the classroom, who cannot attain the mental focus necessary for learning.

At this point let's introduce the notion of the benevolent metals, trace or otherwise, which will counteract such effects. Calcum and zinc are the most vital for any process of *chelation:* that is the remedial process whereby the toxic metals are flushed out of the system, and they are used together with vitamin C. But also there are elements you have never thought about such as selenium, which are only needed in very low levels, but they are needed. Modern agriculture flushes out such trace elements from topsoil and so our diets are liable to be lacking in it.

From a sample of over five hundred hyperactive children living around Oxford, Neil Ward was able to obtain hair and blood samples. These two indicators have very different meanings: blood gives the current level in the

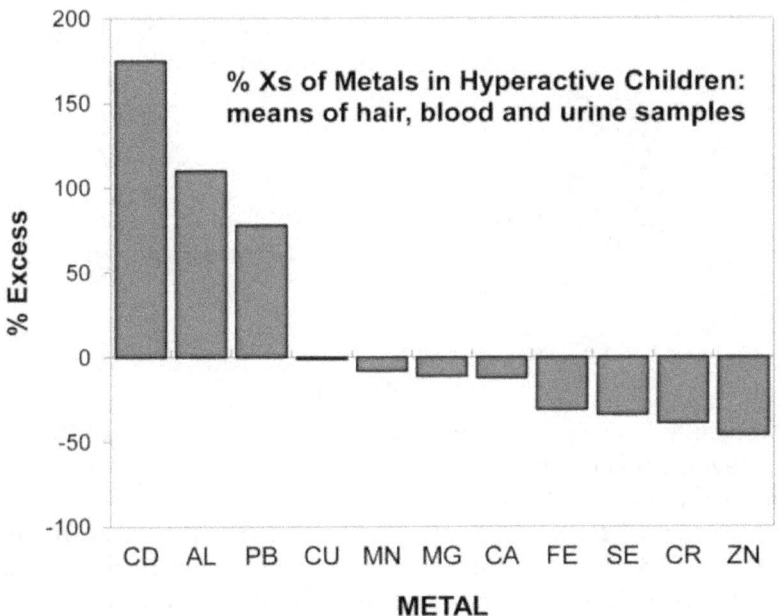

body whereas hair represents a long-term storage. Indeed heavy metals may be expelled by being put into hair. (Isaac Newton was found to have high mercury levels in his hair in a posthumous testing). There are a whole lot of protocols of washing the hair, drying it, where it is sampled etc and let's just take these as read.

I've tried to present the data here in a form that is accessible for health workers etc and does not give a load of numbers that are instantly forgettable and have very little significance. We see here three *neurotoxic* metals: cadmium, lead and aluminium, elevated around 100%, in other words double the levels one would find in the normal population. These three metals are having a terrible effect upon children diagnosed as hyperactive. These youngsters are also short of essential or benevolent metals: iron, selenium, chromium and zinc. Such shortages are in general a function of modern agribusiness whereby crops and animals are force-fed and made to grow too quickly.

Compare that with a similar graph of violent young criminals, who show much higher excessive levels of aluminium and lead. This is a much smaller sample (28, males aged 16-19 years) and not from any criminals in the UK. Neil Ward was fortunate in being able to get hair and blood samples from such incarcerated offenders and I guess it would be impossible to obtain any such data in the UK.

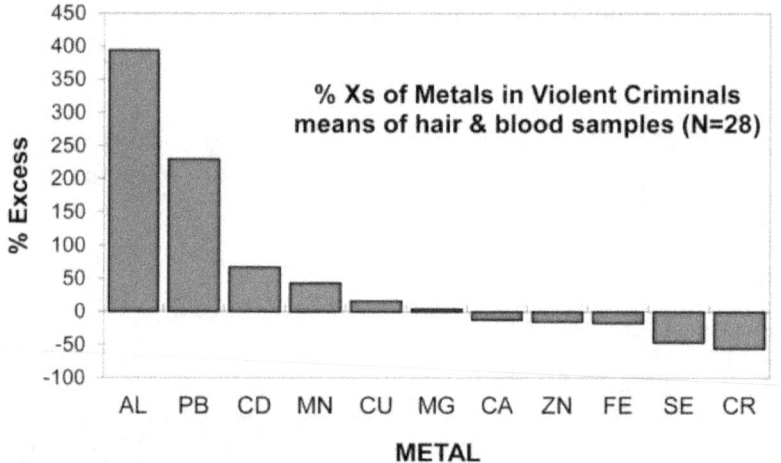

We can only wonder, why does no-one treat violent young criminals by chelation therapy? These toxic-metal levels are going to separate such VYCs from a normal population better than any other factor.

If anyone wanted to do a follow-up UK study, my suggestion would be to take a high and low ability stream in a modern comprehensive class. Or indeed a sample of *Sun*-readers versus say *Guardian* readers. When I was in the Green Party (It was then the Ecology Party, this is really going back) and they were helping me to promote my *Lead on the Brain* book, we were discussing the tragic stunting of national IQ by lead. One guy remarked 'That's how the Sun makes its profits.' I would bet a substantial sum of money that the above sampling suggestion would obtain a significant result. But, as well as that, one would like to see a multi-element sample of his kind done on, say an orchestra, where persons of a harmonious and more tranquil disposition may be found.

I wrote this up in an online article for Biolab[180]. N.B., They are in London and will do hair analysis.

We need to find some way of communicating to the general population in general and pre-conceptual couples in particular that these 'homeopathic' low levels of the toxic metals do really matter. Cadmium comes from cigarette smoke and coca-cola, aluminium from cooking utensils and lead is still generally ambient though less so these days. If you're in a soft water area be more careful because such water will not have the dissolved calcium, which will in some degree protect the body against lead absorption.

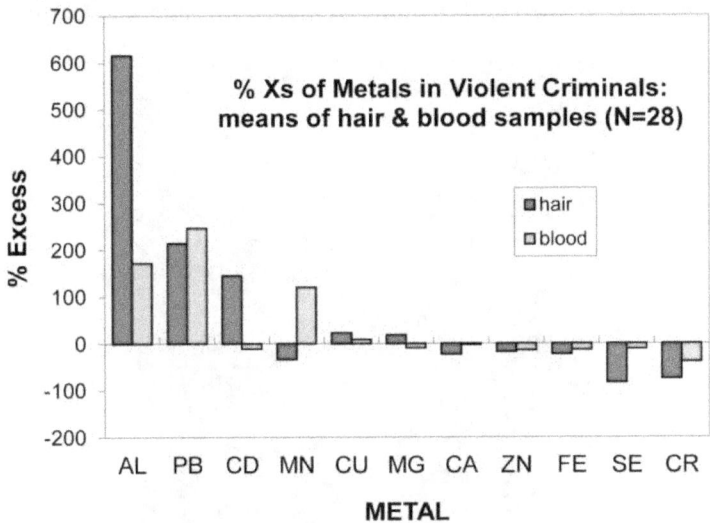

A third graph shows much the same as before only the blood and hair samples are kept separate. This I hope just shows how it is really worth having these two separate samples to average together. Hair stores the toxic metals, or they can be excreted into hair, while blood gives the present-time body level.

Dr William Walsh, addressing the Well Mind Association (US) in 1991, remarked: 'The answer to crime prevention is not in bigger prisons and more stringent penalties but in identifying children and intervening biochemically before their lives are ruined.'

Thirty Years Later

'I will show you fear in a handful of dust'
T.S.Eliott

---

[180] http://www.biolab.co.uk/docs/nkpaper.pdf

My *Lead on the Brain* book was published in 1982 during the 'great British lead debate' – synchronizing with the triple Jupiter-Saturn conjunction – where the government finally decided to remove lead from petrol. It had sort of conceded the point that the neurotoxic effects were working at a much lower level than it had been maintaining. However its main concern seemed to be to avoid litigation against claims that it had allowed child brain damage. I was rather baffled at the way every one seemed to think we could now forget about this unpleasant subject. Massive amounts of lead had been put into the biosphere and it would be having far reaching effects. Were parents just not bothered about permanent child-brain damage? Or did the notion of one single element producing disturbed and delinquent kids, seem like science fiction?

Fast forward thirty years and a US report is arguing for a massive correlation between violent crime and ambient lead levels. As journalist George Montbiot argued in *The Guardian,*

> It seems crazy, but the evidence about lead is stacking up. Behind crimes that have destroyed so many lives, is there a much greater crime?

Globally, it transpires that 'The rise and fall of violent crime during the second half of the 20th century and first years of the 21st were caused, it proposed, not by changes in policing or imprisonment, single parenthood, recession, crack cocaine or the legalisation of abortion, but mainly by ... lead.'

A time lag was found of just over twenty years, between a rise and fall in the exposure of infants to trace quantities of lead, and similar curve for violent crime rates:

> The curve is much the same in all the countries these papers have studied. Lead was withdrawn first from paint and then from petrol at different times in different places (beginning in the 1970s in the US in the case of petrol, and the 1990s in many parts of Europe), yet despite these different times and different circumstances, the pattern is the same: violent crime peaks around 20 years after lead pollution peaks. The crime rates in big and small cities in the US, once wildly different, have now converged, also some 20 years after the phase-out.
>
> One paper found, after 15 variables had been taken into account, a four-fold increase in homicides in US counties with the highest lead pollution. Another discovered that lead levels appeared to explain 90% of the difference in rates of aggravated assault between US cities.

A study in Cincinnati finds that young people prosecuted for delinquency are four times more likely than the general population to have high levels of lead in their bones. A meta-analysis (a study of studies) of 19 papers found no evidence that other factors could explain the correlation between exposure to lead and conduct problems in young people

Lead poisoning in infancy, even at very low levels, impairs the development of those parts of the brain (the anterior and prefrontal cortex) that regulate behaviour and mood. The effect is stronger in boys than in girls. Lead poisoning is associated with attention deficit disorder, impulsiveness, aggression and, according to one paper, psychopathy. Lead is so toxic that it is unsafe at any level.

Let's underscore that: *unsafe at any level.* That means even below 0.1 parts per million in the blood – and remember during the 1980s no-one could even measure reliably down at that level – a negative correlation continues to exist, between measures of IQ and blood lead level.

We sense the Saturnine themes here involved. This does help us to comprehend why people don't seem to want to know about something that is exerting such a baneful influence upon our whole civilization. A couple of decades separate the cause and effect: early childhood lead ingestion and the mentally stunted, violent-prone youth. Also the low levels tend to confuse people: experts will talk about 25 milligrams per decilitre of lead in the blood – what I'm here calling 0.25 parts per million. Law and order is itself a Saturnine issue. Buy nobody in that area is trained to consider that body heavy metal

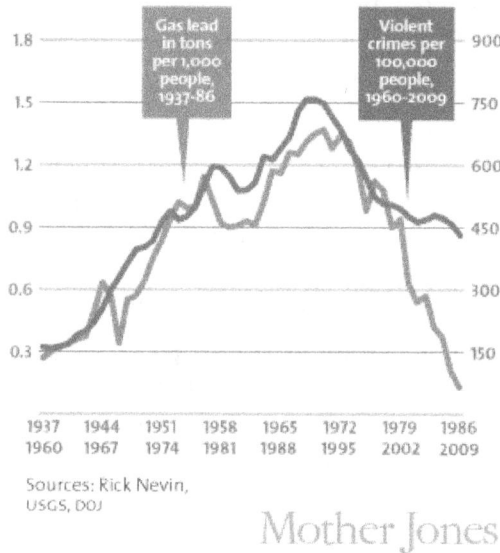

Sources: Rick Nevin, USGS, DOJ

Mother Jones

levels may have a vastly greater effect in producing violent young criminals than any of the sociological effects they like to bang on about, such as broken families.

So this is a *chronic* problem – from *Chronos,* Old Father Time – which extends over generations. Ambient lead remains in city dust for a long time, the rain does not wash it away.

There is a racial difference here, with one 1986 study finding that 18% of white children but 52% of black children in the US had over 20 milligrams per decilitre of lead in their blood; another found that, between 1976 and 1980, black infants were eight times more likely to be carrying the horrendous load of 40mg/dl. This, two papers propose, could explain much of the difference in crime rates between black and white Americans.

In city after city, violent crime peaked in the early '90s and then began a steady and spectacular decline. The hypothesis was put forward, that this was due primarily to tetraethyl lead, from car exhaust. A paper from 2000 showed, that if you add a lag time of 23 years, lead emissions from automobiles explain 90 percent of the variation in violent crime in America. Toddlers who ingested high levels of lead in the '40s and '50s really were more likely to become violent criminals in the '60s, '70s, and '80s.

But, correlation doesn't show causality, people said. So then a study showed that in states where consumption of leaded gasoline declined slowly, crime declined slowly. Where it declined quickly, crime declined quickly. This pattern was found repeatedly in different nations around the world – lead data and crime data for Australia showed a close match. Ditto for Canada. And Great Britain and Finland and France and Italy and New Zealand and West Germany. Every time, the two curves fit each other astonishingly well. The Mother Jones journalist reported, "When I spoke to Nevin about this, I asked him if he had ever found a country that didn't fit the theory. "No," he replied. "Not one.'"

Even lead concentrations at the *neighborhood* level in New Orleans matched up with crime maps supplied by the local police.

This theory helps to explain something we might not have realized even needed explaining, that murder rates were higher in big cities than in towns and small cities. We're so used to this that it seems unsurprising, but it seems likely that big cities have lots of cars in a small area and so also had high densities of atmospheric lead during the postwar era. But as lead levels in gasoline decreased, the differences between big and small cities largely went away. Then, guess what? The difference in murder rates went away too. Today, homicide rates are similar in cities of all sizes. It may be that violent crime isn't an inevitable consequence of being a big city after all.

You might have been surprised that, in the above graphs, mercury did not appear as a harmful toxic metal. 'Mercury's in the clear' Neil Ward told me.

The EPA now says flatly that there is "no demonstrated safe concentration of lead in blood," and it turns out that even levels under 10 µg/dL can reduce IQ by as much as seven points.'

These results have vindicated the argument of my book *Lead on the Brain*. But beyond that, with a modern multi-element atomic absorption spectrophotometry, one can readily obtain say fifteen different metals from a single sample. So there is cause for a guarded sense of optimism. And it's important for governments to forgive themselves for having put far too high 'safety thresholds' in the past. This is an evolving process and cities are going to have ambient toxic-metal levels around in the forseable future. Countries do needs an EPA environmental Protection Agencies which can keep an eye on these matters, which most people don't want to bother with.

What I call New Alchemy involves an awareness of how the concentrations of what a homeopath would call D7 and indeed D8 of these toxic metals really work in the well-being of civilization and the wellness of children. These would be expressed as 1 in 10^7 and 10^8 by weight. D8 is ten parts per billion. Ordinary folk are never going to remember these (I notice), but we do really need some responsible group of eco-experts advising the government who will not get so muddled up.

(Search for, lead crime link gasoline, Montbiot Guardian, Mother Jones, January 2013)

Back in 1980, when the great lead-debate got into full swing, government experts would try to argue that the case against lead in petrol was very much due to one man: Professor Derek Bryce-Smith, chemistry lecturer at Reading University. He was the main author of 'Lead or Health' a detailed rebuttal of the government's view, which came out while I was working at the MRC's air Pollution Research Unit. I came to apprehend that he was in fact correct, and that the department I was working for was very much involved in whitewashing the shocking evidence of child brain damage, due to its too-close link with the petrochemical industry. Upon visiting Bryce-Smith one day in Reading, I summoned up my courage and asked for his birth date.

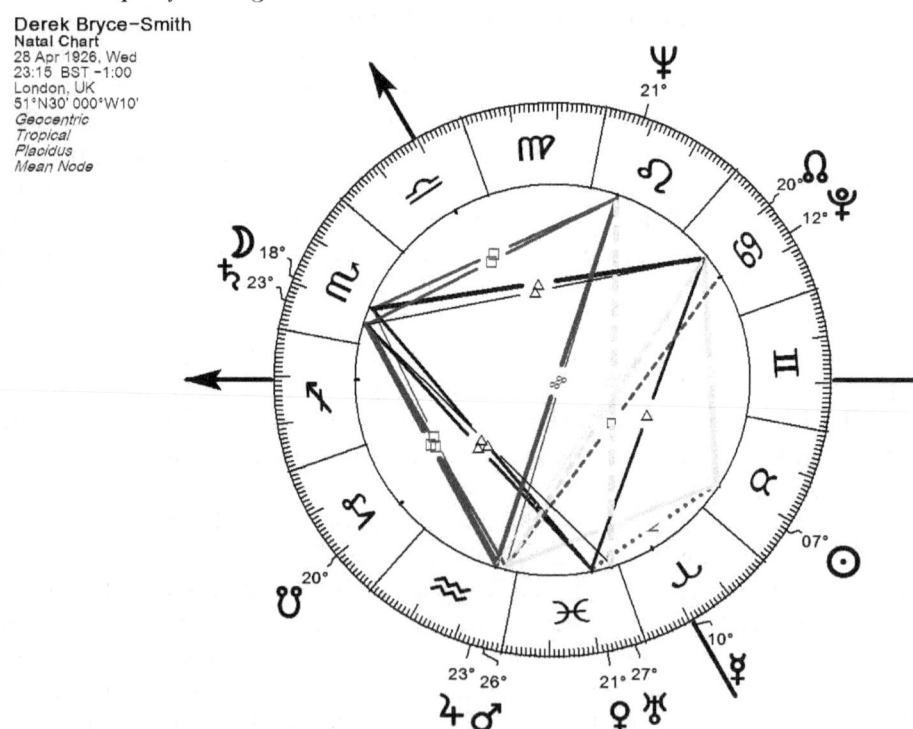

Figure: the chart of Professor Derek-Bryce-Smith, chemistry professor at Reading University, co-author of 'Lead or Health,' showing its great focus upon Saturn (28 Apr 1928, 23.15 BST London).[181]

I doubt whether you will ever see another chart so utterly focused upon Saturn. Everything in the chart is pointing at it! Is that not what gave him his tremendous insight into lead? Any astrologers reading this, may wish to check out how the major, triple Jupiter-Saturn conjunction around 1980 was touching that chart. After all, that's how destiny works, isn't it?

---

[181] There are four quintile aspects here shown, in pale grey, (dividing the circle by five) not relevant to the present treatise, but said to be associated with creative mental activity

2. Cosmic Timing of Gold & Silver Prices

Some time ago, I was sent some years of daily bullion prices of the London Metal Exchange, by Ray Merriman. For the value of gold, I found only its fundamental yearly swing, which I presume is well-known though I have never seen it stated.

This is a 'moving average,' using weekly mean gold-price values over 20 years. It shows how the price reaches its maximum in January-February. When we are most starved of the Sun's rays, the Sun-metal gold is most desired (a 6% swing – probably not worth trading on!) I looked for but could not find a mercury-pattern, of its conjunctions with the Sun.

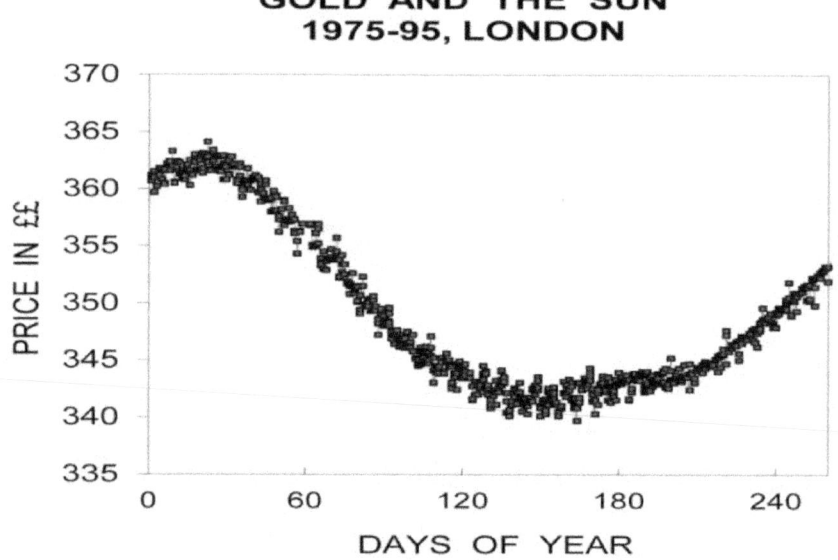

I had four years of silver price data. Plotting this against the Full Moon seemed to give a peak just before the Full moon. The graph shows this plotted as the percentage swing in silver prices through the lunar month, over those four years: it looks like a several percent effect.

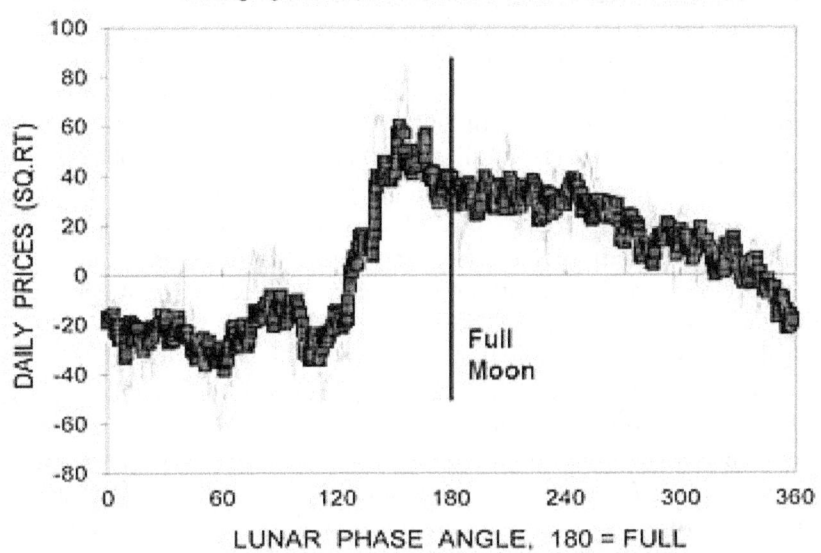

I was also given the daily trading volume of silver. This may be a better value to use, because it does not attract so much the greed and egoism as do silver prices. As before I took square roots of the daily trade volumes, to try and even out the sudden big fluctuations, then I divided all of the lunar months into either high node or low-node. Bear with me while I explain this.

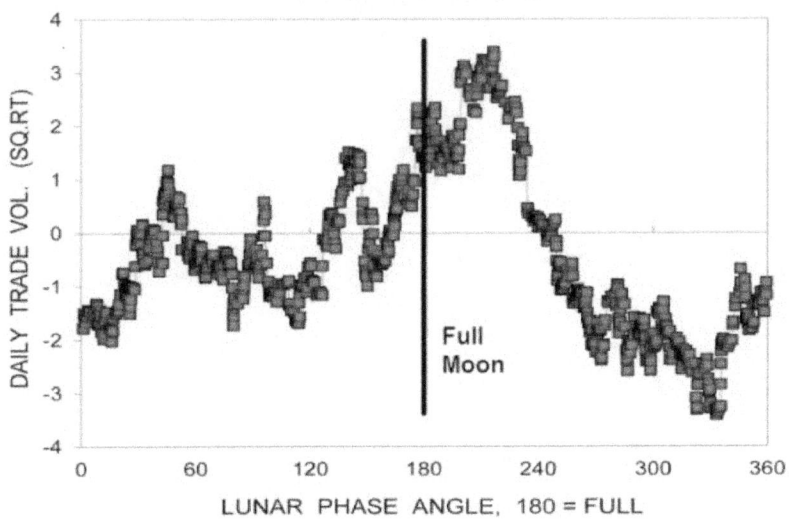

At a Full Moon, the Sun and Moon are opposite each other in the heavens, and the straight line they form may be near to or far from the axis of the lunar nodes. Obviously when these two lines conjoin an eclipse takes place. We are really outside the scope of this book at this point: you may wish to peruse my *Farmer's Moon* which explains this matter. The power of a full Moon will differ according to this cycle. So what I'm here calling 'low node' is (perhaps confusingly) those lunar cycles when the Full Moon is close to the ecliptic i.e. near to its node. As it is then opposite the Sun, we just select that half of the data where the Sun is near (by ecliptic longitude angle) to one of the nodes.

Here it looks as if the main increase in daily trade is happening just after the Full Moons – those of low celestial latitude, i.e., near to the ecliptic. All graphs have used moving averages to smooth the data. These have not been published, as I was (and still am) hoping to try and replicate them with some fresh data.

We should not be surprised if, when Selene's sphere is shining most brightly, silver prices rise and more is traded.

This being an alchemical treatise we are not really concerned with facts and figures, but more with *quality* and with essential being. However, if an estimate of the monthly 'lunar swing' in silver volume trading is desired, it amounted to nearly 4% between the mean values for the third and 4th lunar quarters.

Bibliography

John T. Burns, Cosmic Influences on Humans, Animals and Plants, an Annotated Bibliography 1997

William Cloos, The Living Earth, 1977

Alison Davidson, Metal Power, the soul life of the planets CA 1992 In print)

Geoffrey Dean, Recent advances in Natal Astrology, 1977.

Richard Grossinger The Alchemical Tradition in the late 20th Century 1983

Rudolf Hauschka, The Nature of substance Spirit and Matter 1966, 1969

J. Partington, Vol. I History of Chemistry, Theoretical Background 1970

Elizabeth Kolisko Working of the Stars in Earthly Substance 1928
- Das Silver und der Mond 1929
- Der Jupiter und das Zinn   1932
- Gold and the Sun   1947
- Spirit in Matter 1948
- Saturn und Blei   1952

N.K. Lead on the Brain, 1982
- Astrochemistry, a Study of Metal-Planet Affinities   1984
- The Metal-Planet Relationship CA 1993

Primo Levi, The Periodic Table 1987

Frank McGillion   The Opening Eye 1982

Maurice Nicholl The Chemical Theatre   1980

Wilhelm Pelikan The Secrets of Metals NY   1973

Mellie Uyldert   Metal Magic   1980.

Index

Africa 36, 103
Al-Biruni   117
Alexandria   28
Aldebaran 162
Antimony 123
Aphrodite 51, 54, 73, 88
Aphrodisias 88
Artemis 39
Assyrians   112
Astrochemistry 2
Atomic number 24
Aurora 84, 102
Bacon, Francis 10
BBC Horizon 145
Becker, R 45
Bismuth 75, 123-5
Blood serum 42
Botticelli 88
Boyle, Robert 11, 122
Bradbury, Ray 108
Brain-Bio Centre 91
Brass 125
Bronze 58, 111
Byde, John 93
Celsus   112
Chinese alchemy 182
Chu, Ching-Wu 168
Cinnabar   115
Colloidal silver 45
Contraceptive 49, 92
Cyprus   45, 114

Dante 16
Days of week 22
Dee, John   129
Desbiolles, G. 41, 149
Drummond, M. 142-5
Electrode potential 17, 44,
Elohim   28
Emerald Table   60, 115
Euripides   114
Faraday, M. 52-96, 159
Fast-breeder reactor 154
Faust, Goethe's 109
Fermi, Enrico 160
Flamel, Nicholas 128
Fyfe, Agnes 40, 71, 83, 93
Gauquelin, M. 44,173
Ginsberg, A., 154
Gold monomolecular 104
Gold trading 193
Haephastos 53
Hahnemann, S. 97
Harvey, Charles   141
Heart of Glass 106
Helvetius   130
Heptagons 25
Hermes   60, 85
Hiroshima 164
Holmyard, E.J. 112
Homer   67, 87
Homeopathy 83, 184, 192
Jet plane, Whittle 66

Johnson, Ben 77
Kelley, Edward 129
Kepler Johannes 4, 173
Kolisko, E. 42, 141-6
Laing, R.D. 10
Lehman, Lee 183
Lemery, Nicholas 125
Lerner, Mark 165
Levi, Primo 62,79
Lycurgus cup 106
Lions, white 103
Maier, Michael 176
Manchester computer 168
McGillion, Frank 20, 151
McLean, Adam 72, 142
Merriman, R. 195
Mitchell, John 105
Mithraic religion 34,113
Morton, anaesthesia 72
Nanotechnology 98
Neptune 171
Newton, I. 73,119, 122
Ouroboros 156
Parrish, Maxfield 52
Pelikan, W. 14, 27, 141
Perfection, scale of 114
Pewter 58
Photography 39
Piccardi, 76
Platinum 13, 37, 97, 125
Pliny 47, 105
Pluto node 153
Price, James 133
Purple of Cassius 107
Rasayana 47, 76, 82
Redheads 174
Red mercury 78

Revelation, Book of 105
Royal Society, the 134
Ruby glass 106
Runge, O. 102
Sagan, Carl 9,173
Sanskrit 81
Saturnism 59
Seaborg, Glen 7,156
Septile aspect 21
Seven days of week 23
Silver trading 195
Sleeping Beauty 102
Steigbild 42
Steiner, Rudolf 35, 97,140
Superconductor 169
Thermonuclear heat 109
Thoth 66,115
Tin mine Cornish 57
Tucker L. 104
UFOs and mercury 81
Uranus 52
Uranium 60
Uyldert, Mellie 88
Venetian glass 106
Venus, harmony 178-80
Vettius Valens 103-118
Viking spacecraft 180
Vimana 82
Vitriol of Mars 126
Whittle, Frank 166

www.ingramcontent.com/pod-product-compliance
Lightning Source LLC
Chambersburg PA
CBHW060842170526
45158CB00001B/214